The ESSENTIALS of GROUP THEORY I

Emil G. Milewski, Ph.D.

> This book covers the usual course outline of Group Theory I. For additional topics, see *"THE ESSENTIALS OF GROUP THEORY II"*.

Research and Education Association
61 Ethel Road West
Piscataway, New Jersey 08854

# THE ESSENTIALS OF GROUP THEORY I ®

Copyright © 1989 by Research and Education Association. All rights reserved. No part of this book may be reproduced in any form without permission of the publisher.

Printed in the United States of America

Library of Congress Catalog Card Number 89-61503

International Standard Book Number 0-87891-686-5

ESSENTIALS is a registered trademark of
Research and Education Association, Piscataway, New Jersey 08854

WHAT "THE ESSENTIALS" WILL DO FOR YOU

This book is a review and study guide. It is comprehensive and it is concise.

It helps in preparing for exams, in doing homework, and remains a handy reference source at all times.

It condenses the vast amount of detail characteristic of the subject matter and summarizes the **essentials** of the field.

It will thus save hours of study and preparation time.

The book provides quick access to the important facts, principles, theorems, concepts, and equations in the field.

Materials needed for exams can be reviewed in summary form – eliminating the need to read and re-read many pages of textbook and class notes. The summaries will even tend to bring detail to mind that had been previously read or noted.

This "ESSENTIALS" book has been prepared by an expert in the field, and has been carefully reviewed to assure accuracy and maximum usefulness.

Dr. Max Fogiel
Program Director

CONTENTS

Chapter No.		Page No.
1	**SETS AND MAPPINGS**	1
1.1	Basic Notation. Propositions	1
1.2	Sets. Subsets	5
1.3	Operations on Sets	7
1.4	Cartesian Product. Ordered Pair	12
1.5	Equivalence Relations	13
1.6	Mappings	15
1.7	Composition of Mappings	17
1.8	Algebraic Laws	17
2	**GROUPOIDS AND SEMI-GROUPS**	21
2.1	Groupoids. Commutative Groupoids. Quasi-Group. Semi-Group.	21
2.2	Unit and Inverse Elements	23
2.3	Semi-Group of Mappings	27
3	**GROUPS**	30
3.1	Groups; an Introduction. Definition of Groups	30
3.2	General Products	36
3.3	Subgroups	37
3.4	Permutations. Symmetric Group	38
3.5	Alternating Groups. Even and Odd Permutations. Order of a Group	40
3.6	Groups of Isometries	44
4	**ISOMORPHISMS AND HOMOMORPHISMS**	52
4.1	Isomorphisms	52

4.2	Homomorphisms, Cayley Theorem.	55
4.3	The Subgroup Generated by X	57
5	**CYCLIC GROUPS. COSETS**	**61**
5.1	Cyclic Groups. Definition of Abelian Groups.	61
5.2	Subgroups of Cyclic Groups	66
5.3	Cosets. Index. Theorem of Lagrange.	67
5.4	Normal Subgroups. Center of a Group. Centralizer. Normalizer.	71
5.5	Examples of Groups. Dihedral Groups. Symmetries of the Cube. Quaternion Group	73
6	**HOMOMORPHISMS**	**77**
6.1	The Kernel of a Homomorphism. First Theorem on Homomorphisms.	77
6.2	Factor Groups. Second and Third Theorems on Homomorphisms.	78
7	**THE SYLOW THEOREMS**	**82**
7.1	Theorem of Lagrange	82
7.2	The Sylow Theorems. H-conjugate.	83
7.3	Class Equation of G	87
8	**FINITE p-GROUPS**	**90**
8.1	Construction of Finite p-groups	90
8.2	Groups of Order p, p^2, pq, p^3. Direct Product. Internal Direct Product.	93

CHAPTER 1

SETS AND MAPPINGS

1.1 BASIC NOTATION. PROPOSITIONS

To shorten statements, the notation of symbolic logic is frequently used. If a and b are propositions, then:

$a \vee b$ denotes a or b

$a \wedge b$ denotes a and b

$\neg a$ denotes not a

$a \Rightarrow b$ denotes a implies b

$a \Leftrightarrow b$ denotes a is logically equivalent to b

The proposition

$$a \vee b$$

where a and b are propositions, is true if at least one of the components is a true proposition. Thus

$$0 \vee 0 \equiv 0 \qquad 1 \vee 0 \equiv 1$$
$$0 \vee 1 \equiv 1 \qquad 1 \vee 1 \equiv 1$$

The proposition

$$a \wedge b$$

is true if both factors are true propositions.

Note that the equivalence

$$a \equiv b$$

holds if and only if a and b have the same logical value. Operations $\vee$ and $\wedge$ are commutative and associative, i.e.,

$$a \vee b \equiv b \vee a$$
$$a \wedge b \equiv b \wedge a$$
$$a \vee (b \vee c) \equiv (a \vee b) \vee c$$
$$a \wedge (b \wedge c) \equiv (a \wedge b) \wedge c$$

The distributive law holds

$$a \wedge (b \vee c) \equiv (a \wedge b) \vee (a \wedge c)$$

The operation of negation $\rceil a$ sometimes is denoted by $\sim a$ or a'.

The negation of a true proposition is a false proposition and, conversely, the negation of a false proposition is a true proposition.

$$\neg 1 \equiv 0$$
$$\neg 0 \equiv 1$$

The law of double negation holds

$$\neg\neg a \equiv a$$

Here are the two fundamental theorems of Aristotelian logic

$$a \vee \neg a \equiv 1$$
$$a \wedge \neg a \equiv 0$$

De Morgan's laws state that

$$\neg(a \wedge b) \equiv \neg a \vee \neg b$$
$$\neg(a \vee b) \equiv \neg a \wedge \neg b$$

From these laws it is easy to see that operation $\vee$ can be defined with the aid of the $\neg$ operation and the $\wedge$ operation as follows

$$a \vee b \equiv \neg(\neg a \wedge \neg b)$$

Similarly for the $\wedge$ operation we have

$$a \wedge b \equiv \neg(\neg a \vee \neg b)$$

Implication $a \Rightarrow b$ is defined as follows

$$(a \Rightarrow b) \equiv (\neg a \vee b)$$

$a \Rightarrow b$ is read: the proposition a implies the proposition b, or: if a, then b.

We have

$$(0 \Rightarrow 0) \equiv 1$$
$$(0 \Rightarrow 1) \equiv 1 \qquad (1 \Rightarrow 0) \equiv 0$$
$$(1 \Rightarrow 1) \equiv 1$$

Implication has properties similar to deduction.

We have

if $a \Rightarrow b$ and $b \Rightarrow a$ then $a \equiv b$.

Two important laws hold: the law of transitivity of implication (or the syllogism law)

if $a \Rightarrow b$ and $b \Rightarrow c$ then $a \Rightarrow c$

and the law of contraposition (on which the proof by "reductio ad absurdum" is based)

if $\rceil b \Rightarrow \rceil a$ then $a \Rightarrow b$.

It is easy to verify that if $\rceil a \Rightarrow b$ for each b, then a is a true proposition (law of Clausius).

Now we shall introduce the quantifiers.

The quantifier "there exists" is denoted by $\exists$, and the quantifier "for every" is denoted by $\forall$.

EXAMPLE

The assertion: "for each x there exists a y such that for each z,

$a(x, y, z)$ is true" can be written as

$$\forall x \; \exists y \; \forall z : a(x, y, z)$$

1.2 SETS. SUBSETS

At this stage a set is a collection of objects. By

$$a \in A$$

we denote that a is an element of the set A. We shall denote elements by lower-case letters and sets by capital Latin letters. If a is not an element of A we write

$$a \notin A$$

and read this as "a does not belong to A." The most frequently used sets are denoted by

P the set of positive integers 1, 2, 3, ...

N the set of nonnegative integers 0, 1, 2, ... $W = \{0,1,2,...\}$

Z the set of all integers

Q the set of rational numbers

R the set of real numbers

C the set of complex numbers

Often a set is defined by listing its elements.

EXAMPLE

A set consisting of only two elements 1 and 2 is written as

$$\{1, 2\}.$$

EXAMPLE

The set P of positive integers can be written as

$$\{1, 2, 3, \ldots\}.$$

A set can also be described in terms of a property which singles out its elements. Then we write

$$\{x \mid x \text{ has the property } p\}$$

for the set of all those elements x which have the property p.

EXAMPLE

The set of real numbers can be written as follows

$$\{x \mid x \in R\}.$$

EXAMPLE

The set of rational numbers can be written as

$$\{x \mid x = \frac{a}{b}, \text{where } a, b \in Z \text{ and } b \neq 0\}.$$

When two sets A and B are equal, we write

$$A = B,$$

if every element of A belongs to B and every element of B belongs to A.

EXAMPLE

$$\{1, 2, 3\} = \{2, 1, 3\} = \{2, 2, 1, 1, 3\}$$

If every element of the set A is also an element of the set B, we say A is a subset of B and denote it by

$$A \subseteq B.$$

If $A \subseteq B$ and $A \neq B$ then A is a proper subset of B, we write

$$A \subset B.$$

Often the equality of sets is established with the help of this theorem.

THEOREM 1

Set A is equal to set B if and only if $A \subseteq B$ and $B \subseteq A$.

1.3 OPERATIONS ON SETS

Let A and B be sets. The union of A and B, written

$$A \cup B$$

is defined as the set whose elements are either in A or in B (or in both A and B). Thus

$$x \in A \cup B \equiv (x \in A) \lor (x \in B).$$

EXAMPLE

$$A = \{1, 2, 3\} \quad B = \{2, x\}$$
$$A \cup B = \{1, 2, 3, x\}$$

$A \cup B$ is the smallest set containing A and B.

The intersection of sets A and B, denoted by

$$A \cap B$$

is the set consisting of those elements which belong simultaneously to A and to B.

$$x \in A \cap B \equiv (x \in A) \wedge (x \in B).$$

The empty set, denoted by ϕ, is the set with no elements. The empty set ϕ is a subset of every set.

$$\phi \subseteq X$$

for any set X.

Two sets are called disjoint if they have an empty intersection, i.e.,

$$A \cap B = \phi.$$

The difference of sets A and B, denoted by

$$A - B$$

is the set of those elements which belong to the set A but do not belong to the set B.

$$x \in A - B \equiv (x \in A) \wedge (x \notin B).$$

For any set A

$$A - A = \phi,$$

Applying Venn diagrams, we can illustrate the above definitions.

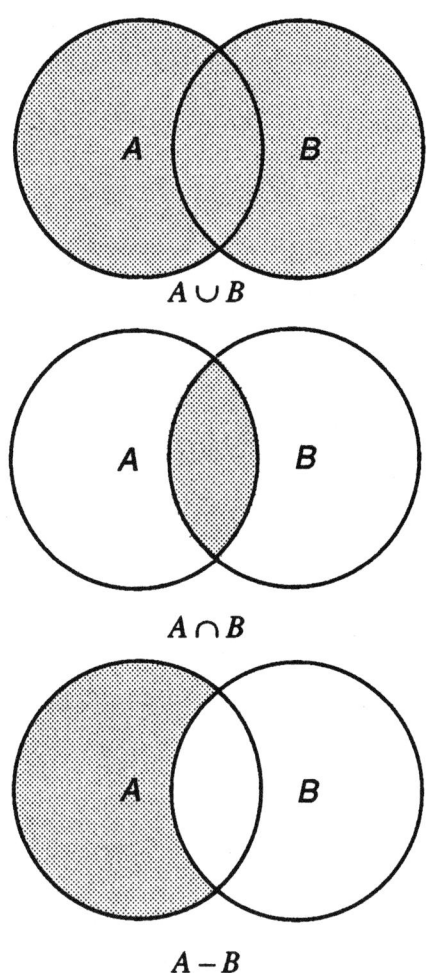

$A \cup B$

$A \cap B$

$A - B$

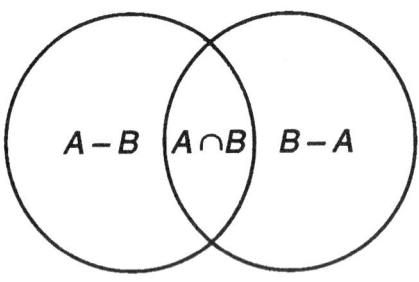

The operations of union and intersection of sets are commutative, i.e.
$$A \cup B = B \cup A$$
$$A \cap B = B \cap A$$
and associative
$$A \cup (B \cup C) = (A \cup B) \cup C$$
$$A \cap (B \cap C) = (A \cap B) \cap C.$$

The distributive law holds
$$A \cap (B \cup C) = (A \cap B) \cup (A \cap C)$$
also
$$A \cup (B \cap C) = (A \cup B) \cap (A \cup C).$$

Note that
$$A \cup A = A \cap A = A.$$

Finally we shall define the complement of a set.

As a rule, we assume that all the sets under consideration are subsets of some fixed set, called the space. Let us denote the space by X.

The complement of the set A with respect to the given space X is defined by

$$A^c = X - A.$$

The complement of A is denoted by A^c (or by $\sim A$ or by CA). We have

$$x \in A^c \equiv X \notin A.$$

The following formulas hold

$$\phi^c = X$$
$$X^c = \phi$$
$$A^{cc} = A$$
$$A \cup A^c = X, \quad A \cap A^c = \phi$$

and

$$A - B = A \cap B^c.$$

THEOREM 2

If A and B are subsets of some universal space X, then

$$(A \cup B)^c = A^c \cap B^c$$
$$(A \cap B)^c = A^c \cup B^c.$$

The following equivalence is frequently applied

$$(A \subseteq B) \equiv (A \cap B^c = \phi).$$

1.4 CARTESIAN PRODUCT

An ordered pair

$$(a, b)$$

means: "a in the first place, b in the second place."

The pair (a, b) is distinct from (b, a) unless $a = b$. An ordered pair can be defined in the following way:

$$(a, b) = \{\, \{a\}, \{a, b\} \,\}.$$

The pairs (a, b) and (c, d) are equal only when $a = c$ and $b = d$, i.e.

$$(\,(a, b) = (c, d)\,) \Leftrightarrow (\,(a = c) \wedge (b = d)\,).$$

The cartesian product of the sets A and B is the set of all ordered pairs (a, b) where $a \in A$ and $b \in B$. We denote this set by $A \times B$ and therefore

$$A \times B = \{(a, b) \mid a \in A \wedge b \in B \}.$$

The product of n sets $A_1, \ldots, A_n$ is denoted by

$$\mathop{\mathrm{X}}_{i=1}^{n} A_i = A_1 \times A_2 \times \ldots \times A_n$$

THEOREM 3

Let A, B, and C be sets. Then

$$A \times (B \cup C) = (A \times B) \cup (A \times C)$$
$$A \times (B \cap C) = (A \times B) \cap (A \times C)$$

$$A \times (B - C) = (A \times B) - (A \times C)$$

1.5 EQUIVALENCE RELATIONS

DEFINITION OF RELATION

A subset R of the product

$$X \times Y$$

is called a relation,

$$R \subseteq X \times Y$$

x is said to be in relation R with y, when $(x, y) \in R$, we write

$$x \, R \, y.$$

DEFINITION OF REFLEXIVE, SYMMETRIC, AND TRANSITIVE RELATIONS

A relation R is called reflexive if

$$x \, R \, x \text{ for each } x.$$

R is symmetric if

$$x \, R \, y \Rightarrow y \, R \, x$$

R is transitive if

$$x \, R \, y \text{ and } y \, R \, z \Rightarrow x \, R \, z$$

DEFINITION OF EQUIVALENCE RELATION

Relation R is called an equivalence relation if R is reflexive, symmetric, and transitive.

EXAMPLE

A partition of a set X into disjoint classes

$$X = \bigcup_{i \in I} A_i$$

defines a relation $R \subseteq X \times X$, that is $x \, R \, y$ when there exists $i \in I$ such that

$$x, y \in A_i,$$

i.e. when x and y belong to the same class A_i. Relation R is an equivalence relation.

If $a \, R \, b$, we say that a and b are equivalent.

Let A be any set and $R \subseteq A \times A$ be an equivalence relation. For $a \in A$, the equivalence class of a, denoted by $[a]$ is defined as

$$[a] = \{ \, b \in A \mid b \, R \, a \, \}.$$

THEOREM 4

Any equivalence relation in X defines a partition of X into (disjoint) classes. Each class is of the form

$$[a] = \{ \, b \in X \mid a \, R \, b \, \}.$$

The set of all classes $[x]$ is called the quotient space of X by R and is written

The map

$$p_X : X \to \frac{X}{R}$$

defined by $x \to [x]$ is called the projection of X onto $\frac{X}{R}$. Each element x of an R-equivalence class $[x]$ is called a representative of $[x]$.

1.6 MAPPINGS

DEFINITION

Let X and Y be two sets. A mapping (map)

$$f : X \to Y$$

is a subset, $f \subseteq X \times Y$, such that for each $x \in X$, there is one, and only one, $y \in Y$ satisfying

$$(x, y) \in f.$$

Instead of $(x, y) \in f$ we write

$$y = f(x).$$

EXAMPLE

The mapping $x \to x$ of X onto itself is called the identity map of X.

DEFINITION

Let $f : X \to Y$. Then:

1. The image of $A \subseteq X$ in Y under f is

$$f(A) = \{ f(x) \mid x \in A \}.$$

2. The inverse image of $B \subseteq Y$ in X is

$$f^{-1}(B) = \{ x \in A \mid f(x) \in B \}.$$

DEFINITION

Let A be any set. Its power set $P(A)$ is the set of all subsets of A.

The mapping $f: X \to Y$ induces two mappings:

1. $f: P(X) \to P(Y)$ by $A \to f(A)$.

2. $f^{-1}: P(Y) \to P(X)$ by $B \to f^{-1}(B)$.

THEOREM 5

$$f^{-1}(\bigcup_\alpha B_\alpha) = \bigcup_\alpha f^{-1}(B_\alpha)$$

$$f^{-1}(\bigcap_\alpha B_\alpha) = \bigcap_\alpha f^{-1}(B_\alpha)$$

$$f^{-1}(B_1 - B_2) = f^{-1}(B_1) - f^{-1}(B_2)$$

THEOREM 6

$$f(\bigcup_\alpha A_\alpha) = \bigcup_\alpha f(A_\alpha)$$

$$f(\bigcap_\alpha A_\alpha) = \bigcap_\alpha f(A_\alpha)$$

An injection (one-to-one mapping) is such that

$$\forall\, x, x_1 \in X\, (f(x) = f(x_1)) \Rightarrow (x = x_1)$$

A surjection (onto mapping) is a mapping of f such that

$$f(X) = Y$$

A bijection is a mapping which is both a surjection and an injection.

1.7 COMPOSITION OF MAPPINGS

Given $f: X \to Y$ and $g: Y \to Z$, their composition

$$g \circ f : X \to Z$$

is defined as the mapping

$$g \circ f : x \to g(f(x))$$

THEOREM 7

Let $f: X \to Y$ and $g: Y \to Z$. Then

$$(g \circ f)^{-1} = f^{-1} \circ g^{-1}$$

1.8 ALGEBRAIC LAWS

Let $S(a, b, c, ...)$ be a system whose elements are designated by letters $a, b, c \ldots$. Operations of addition and multiplication are defined on S. Here is the list of basic laws. Which of these laws hold depends on what algebraic structure we investigate.

CLOSURE LAWS

Addition is well defined.

Multiplication is well defined.

That means that for every ordered pair (a, b) of elements $a, b \in S$, $a + b = c$ exists and is a unique element of S (or $ab = c$ exists and is a unique element of S).

ASSOCIATIVE LAWS

$(a + b) + c = a + (b + c)$

$(ab)c = a(bc)$

COMMUTATIVE LAWS

$a + b = b + a$

$ab = ba$

EXISTENCE OF ZERO AND UNITY

0 exists such that

$$a + 0 = 0 + a = a$$

for all $a \in S$.

1 exists such that

$$a \cdot 1 = 1 \cdot a = a$$

for all $a \in S$.

NEGATIVES AND INVERSES

For every $a \in S, -a \in S$ exists such that

$$a + (-a) = (-a) + a = 0.$$

For every $a \in S$ such that $a \neq 0$, $a^{-1} \in S$ exists such that

$$a(a^{-1}) = (a^{-1})a = 1.$$

DISTRIBUTIVE LAWS

$a(b + c) = ab + ac$

$(b + c)a = ba + ca$

DEFINITION OF A FIELD

A system which satisfies all the above laws is called a field.

If for every choice of elements $a_1, a_2, ..., a_n \in S$, the operation

$$f(a_1, a_2, ..., a_n)$$

is defined and unique, we say that the operation f is well defined and that the set S is closed with respect to the operation f.

DEFINITION OF BINARY OPERATIONS

Any mapping h of $A \times A$ into A, where A is any non-empty set, is called a binary operation in A.

EXAMPLE

Let R be the set of real numbers. Then addition is a binary operation in R

$$+ : R \times R \to R$$

$$R \times R \ni (x, y) \to x + y \in R$$

Similarly, multiplication is a binary operation.

The multiplication table is a convenient way of defining a binary operation in a finite set A

Suppose $A = \{1, 2, 3\}$ and the binary operation is defined as follows:

$$(1, 1) \to 2, \quad (1, 2) \to 2, \quad (1,3) \to 3$$
$$(2, 1) \to 1, \quad (2, 2) \to 3, \quad (2,3) \to 3$$
$$(3, 1) \to 1, \quad (3, 2) \to 3, \quad (3,3) \to 2$$

The corresponding multiplication table has the form

	1	2	3
1	2	2	3
2	1	3	3
3	1	3	2

CHAPTER 2

GROUPOIDS AND SEMI-GROUPS

2.1 GROUPOIDS

DEFINITION OF GROUPOIDS

A non-empty set G with a well-defined binary operation is called a groupoid.

Usually a groupoid is denoted by $(G, \cdot)$ or $(G, +)$.

EXAMPLE

Let $G = \{x, y, z\}$ be a set with a binary operation defined by

	x	y	z
x	x	y	y
y	y	x	y
z	z	y	x

Denoting the operation by "o" we obtain a groupoid (G, o).

EXAMPLE

The Set Z of integers with the operation of addition is a groupoid.

Two groupoids (G_1, f_1) and (G_2, f_2) are equal if and only if

$$G_1 = G_2$$

and the binary operations are the same.

DEFINITION OF A COMMUTATIVE GROUPOID

A groupoid $(G, +)$ is said to be commutative or abelian if for all $a, b \in G$

$$a + b = b + a.$$

A commutative groupoid is also called abelian.

DEFINITION OF A QUASI-GROUP

A quasi-group Q is a system in which a binary operation $a \cdot b$ is defined in such a way that, in

$$a \cdot b = c$$

any two of a, b, c determine the third uniquely.

DEFINITION OF A SEMI-GROUP

A semi-group is a system of elements with a binary operation which is associative, i.e. for any a, b, c

$$a(bc) = (ab)c.$$

EXAMPLE

A system of elements $\{1, 2, 3\}$ with a binary operation defined by

	1	2	3
1	2	1	2
2	1	1	3
3	2	3	3

is obviously a groupoid. It is also an abelian groupoid. But it is not a semi-group because

$$(2 \cdot 2) \cdot 3 \neq 2 \cdot (2 \cdot 3)$$

In most cases the multiplicative notation ab will apply to a groupoid and the additive notation $a + b$ will apply to a commutative groupoid.

The number of elements of a groupoid $(G, \cdot)$ is designated by $|G|$ and is called the order of a groupoid.

2.2 THE UNIT AND THE INVERSE ELEMENTS

Let $(G, \cdot)$ be a groupoid. An element $e \in G$ is called the unit element (or the identity element) of G if

$$ea = ae = a$$

for every $a \in G$.

EXAMPLE

1 is the unit element of the multiplication groupoid Z of integers.

EXAMPLE

Let $\{a, b, c\}$ be the groupoid with the multiplicative table

	a	b	c
a	b	a	c
b	a	b	c
c	a	c	a

b is the unit element because

$$ab = ba = a$$
$$bb = b$$
$$cb = bc = c$$

THEOREM 8

A groupoid can have at most one unit element.

Suppose a groupoid G has two unit elements e and e'. Since e is the unit element

$$e\,e' = e'$$

and since e' is the unit element

$$e\,e' = e'e = e.$$

Thus

$$e\,e' = e' = e.$$

EXAMPLE

Let C be the groupoid of complex numbers with the operation of multiplication. Then

$$1 + 0i$$

is the unit element.

$$(a + bi)(1 + 0i) = a + bi.$$

DEFINITION OF INVERSE

If G is a groupoid with the unit 1 then an inverse of $a \in G$ is an element of $b \in G$ such that

$$ab = ba = 1$$

Note that the inverse of an element of G is not uniquely defined.

EXAMPLE

Let $G = \{1, 2, 3\}$ be the groupoid with the multiplication operation

	1	2	3
1	1	2	3
2	2	1	1
3	3	1	2

Here 1 is the unit element of G. Furthermore

$$2 \cdot 2 = 1$$

and

$$2 \cdot 3 = 3 \cdot 2 = 1.$$

Hence 2 has two inverse elements 2 and 3.

What additional properties should a groupoid with the unit element possess to insure the uniqueness of the inverse element?

The next theorem offers the answer.

THEOREM 9

Let G be a semi-group with the unit element 1. Then the inverse of any element $a \in G$, if it exists, is uniquely defined.

Suppose b and b' are inverses of $a \in G$, i.e.

$$ba = ab = 1$$

and

$$b'a = ab' = 1$$

We have

$$b = b \cdot 1 = b(ab') = (ba)b'$$
$$= 1b' = b'$$

We shall introduce the following helpful notation.

Let G be a semigroup with the multiplicative operation, then the inverse of an element $a \in G$ will be denoted by

$$a^{-1}.$$

Suppose $a, b \in G$ have both the inverse elements. Then ab has the inverse element, which is

$$(ab)^{-1} = b^{-1}a^{-1}.$$

Indeed

$$(ab)(b^{-1}a^{-1}) = a(b\, b^{-1})a^{-1}$$
$$= aa^{-1} = 1$$

2.3 THE SEMI-GROUP OF MAPPINGS

Let X be a given non-empty set. By M_X we denote the set of all mappings of X into itself. We shall define the binary operation on M_X to be the composition of mappings, i.e.

$$\text{if} \quad f \in M_X \quad \text{and} \quad g \in M_X$$

then $f \circ g$ is the mapping defined by

$$x (f \circ g) = (xf) g \quad \text{for all } x \in X.$$

THEOREM 10

If X is a non-empty set, then M_X is a semigroup with a unit element.

Operation "o" on M_X is a well-defined binary operation (it is a composition of mappings). Hence M_X is a groupoid. We shall show that M_X is a semi-group. Let

$$f, g, h \in M_X$$

and let $x \in X$. Then

$$x(f(gh)) = (xf)(gh) = ((xf)g)h$$

$$= (x(fg))h = x((fg)h).$$

Thus

$$f(gh) = (fg)h$$

and M_X is a semigroup. We define the identity function as

$$1_X : X \to X$$

$$1_X x = x \quad \text{for all } x \in X$$

If $f \in M_X$, then

$$x(1_X f) = (x\, 1_X)f = xf =$$

$$= (xf)1_X = x(f\, 1_X)$$

Thus
$$1_X f = f 1_X = f$$
for all $f \in M_X$. 1_X is the unit element of M_X.

Observe that not every element of M_X has to have an inverse.

EXAMPLE

Let $X = \{1, 2\}$ and $f \in M_X$ be such that

$$1f = 1, \text{ and } 2f = 1.$$

Then f has no inverse.

THEOREM 11

An element of M_X has an inverse if and only if it is one-to-one and onto.

Let X be a finite set
$$X = \{a_1, ..., a_n\}$$
Any mapping
$$f : X \to X$$
that is any element $f \in M_X$ will be denoted as follows
$$f = \begin{pmatrix} a_1 & ... & a_n \\ a_1 f & ... & a_n f \end{pmatrix}$$

Below an element $a_i \in X$ we place its image $a_i f$ under mapping f.

CHAPTER 3

GROUPS

3.1 GROUPS — INTRODUCTION

We can define a group as a semigroup with an identity, and such that every element has an inverse. We repeat the definition of a group in detail.

DEFINITION OF A GROUP

A group G is a set of elements and a binary operation which we will call "product" such that

1. **Closure Law.** For every ordered pair (a, b) the product ab exists and is a unique element of G.

2. **Associative Law.** For any $a, b, c, \in G$

    $$a(bc) = (ab)c \;.$$

3. **Existence of Unit.** An element $1 \in G$ exists such that for every $a \in G$

    $$a\,1 = 1\,a = a \;.$$

4. **Existence of Inverse.** For every $a \in G$ there exists $a^{-1} \in G$ such that

$$a^{-1} a = a a^{-1} = 1$$

Note that the above definition can be reduced to the shorter yet equivalent version.

Points 3 and 4 can be replaced by

3.* An element $1 \in G$ exists such that for every $a \in G$

$$1 a = a$$

4.* For every $a \in G$ there exists $b \in G$ such that

$$ba = 1$$

3.* and 4.* imply 3. and 4. Indeed, let $a \in G$ then

$$ba = 1 \quad \text{and} \quad cb = 1$$

by 4.* We have

$$ab = 1(ab) = (cb)(ab)$$
$$= c(b(ab)) = c((ba)b)$$
$$= c(1b) = cb = 1.$$

Hence, for every $a \in G$ there exists $b \in G$ such that

$$ba = 1 \quad \text{and} \quad ab = 1.$$

b will be denoted by a^{-1}.

Similarly

$$a = 1a = (aa^{-1})a$$
$$= a(a^{-1}a) = a1.$$

In a group, the unit 1 and an inverse a^{-1} are uniquely defined. We quote another equivalent definition of a group below.

DEFINITION OF A GROUP

A group G is a set consisting of elements such that:

1. For every ordered pair (a, b) of elements $a, b \in G$ a binary product ab is defined such that ab is a unique element of G.

2. To every element $a \in G$ there corresponds a unique element $a^{-1} \in G$.

3. For any choice a, b, c of elements of G

    $$(ab)c = a(bc).$$

4. For any choice of elements a, b, c of G

    $$a^{-1}(ab) = b = (ba)\, a^{-1}.$$

It is easy to show that both definitions are equivalent.

EXAMPLE

The additive group of integers $(Z, +)$.

Let Z be the set of all integers and "+" be the binary operation of addition in Z.

For every $n \in Z$

$$n + 0 = 0 + n = n.$$

Hence 0 is the unit (identity) element of $(Z, +)$. For any $a, b, c \in Z$

$$a + (b + c) = (a + b) + c$$

If $n \in Z$, then $-n \in Z$ and

$$n + (-n) = (-n) + n = 0$$

i.e. $-n$ is an inverse of n in $(Z, +)$.

EXAMPLE

We examine the multiplicative group of nonzero rationals. Q^* is the set of nonzero rational numbers and · is the binary operation of multiplication. 1 is the identify of $(G^*, \cdot)$

$$1 \cdot x = x \cdot 1 = x \text{ for all } x \in G^*.$$

If $a, b, c \in Q^*$, then

$$(ab)c = a(bc).$$

If $x \in Q^*$, then $\frac{1}{x} \in Q^*$ and

$$x \cdot \frac{1}{x} = \frac{1}{x} \cdot x = 1$$

Every element of Q^* has an inverse.

EXAMPLE

We examine the multiplicative group of nonzero complex numbers. Let C^* be the set of all nonzero complex numbers.

$$C^* = \{a + bi \mid a, b \in R, a^2 + b^2 > 0\}.$$

Remember that $i^2 = -1$.

Multiplication of complex numbers is defined as follows:

$$(a + ib)(c + id) = (ac - bd) + i(ad + bc)$$

The product is well-defined because

$$(ac - bd) + i(ad + bc)$$

is a unique element of C^*,

$$(ac - bd)^2 + (ad + bc)^2 > 0.$$

The unit element of $(C^*, \cdot)$ is $1 + 0i = 1$

$$(a + bi) \cdot 1 = 1 \cdot (a + bi) = a + bi.$$

It is easy to verify that the operation of multiplication is associative. Suppose

$$a + bi \in C^*.$$

The inverse element of $a + bi$ is

$$\frac{a}{a^2 + b^2} - i \frac{b}{a^2 + b^2} \in C^*$$

Indeed

$$(a + bi)\left(\frac{a}{a^2 + b^2} - i \frac{b}{a^2 + b^2}\right) =$$

$$= \left(\frac{a}{a^2 + b^2} - i \frac{b}{a^2 + b^2}\right)(a + bi) = 1$$

Thus $(C^*, \cdot)$ is a group.

EXAMPLE

Let $G = \{1, 2\}$ be a set with a binary operation defined by

	1	2
1	1	2
2	2	1

G is a group with the unit element 1. The inverse of 1 is 1, the inverse of 2 is 2

THEOREM 12

If $(G, \cdot)$ is a group, then the identity of G is unique. If $a \in G$, then a has the unique inverse $a^{-1} \in G$. Further

$$(a^{-1})^{-1} = a.$$

THEOREM 13

If $(G, \cdot)$ is a group and $a, b, c \in G$ such that

$$ab = ac,$$

then $b = c$.

If $a, b, c \in G$ such that

$$ba = ca,$$

then $b = c$.

THEOREM 14

If $a, b \in G$, then there exists a unique element, x of G, such that

$$a\,x = b$$

and a unique element, $y \in G$, such that

$$y\,a = b.$$

THEOREM 15

The inverse of a product is the product of the inverses in the reverse order. That is, if

$$a, b, c \in G, \text{ then}$$
$$(ab)^{-1} = b^{-1}a^{-1}$$
$$(abc)^{-1} = c^{-1}b^{-1}a^{-1}$$

3.2 GENERAL PRODUCTS

The associative law states that for all $a_1, a_2, a_3 \in G$

$$(a_1\,a_2)\,a_3 = a_1\,(a_2\,a_3)\,.$$

Thus $a_1\,a_2\,a_3$ is independent of the placement of parentheses.

An important consequence of the associative law is the generalized associative law.

All ways of bracketing an ordered sequence $a_1\,a_2 \ldots a_n$ to give it a value by calculating a succession of binary products, yield the same value.

EXAMPLE

$(a_1 a_2)(a_3 a_4) = (a_1 (a_2 a_3)) a_4 = a_1 a_2 a_3 a_4$

3.3 SUBGROUPS

DEFINITION OF SUBGROUPS

A nonempty set, H, of elements of a group, G, is called a subgroup of G if H is itself a group with respect to the operation already defined on G.

EXAMPLE

If e is the identity of G, then both G and $\{e\}$ are subgroups of G. We call these two subgroups the trivial subgroups.

THEOREM 16

Let $(G, \cdot)$ be a group. A non-empty subset H of G is a subgroup if both conditions hold:

1. If $h_1, h_2 \in H$, then $h_1 h_2 \in H$.

2. If $h \in H$, then $h^{-1} \in H$.

THEOREM 17

Let $(G, \cdot)$ be a group. A nonempty subset H of G is a subgroup of G if and only if for all $a, b \in H$

$$ab^{-1} \in H.$$

EXAMPLE

Let $(Q, +)$ be the additive group of rationals and Z be the set of all integers. If $a, b \in Z$, then

$$a + (-b) = a - b \in Z.$$

Hence, Z is a subgroup of Q.

THEOREM 18

The intersection of two subgroups is a subgroup, but the union of two subgroups is not necessarily a subgroup.

3.4 PERMUTATIONS AND THE SYMMETRIC GROUP

DEFINITION OF A PERMUTATION AND A SYMMETRIC GROUP

A one-to-one mapping of a set onto itself is called a permutation of the set. The group of all permutations of the set is called the symmetric group on the set.

EXAMPLE

Set $\{1, 2\}$ has two permutations

$$(1, 2) \quad \text{and} \quad (2, 1).$$

If t_1 and t_2 are two permutations, then we define $t_1 t_2$ to be the composition of the mappings t_1 and t_2. The identity mapping $t_0 : X \to X$ is the identity element of the symmetric group. The symmetric group on X is denoted by S_x. If $X = \{1, 2, ..., n\}$, we write $S_x = S_n$. S_n is called the symmetric group of degree n.

EXAMPLE

The elements of S_3 are

$$t_0 = \begin{pmatrix} 1 & 2 & 3 \\ 1 & 2 & 3 \end{pmatrix} \quad t_1 = \begin{pmatrix} 1 & 2 & 3 \\ 1 & 3 & 2 \end{pmatrix} \quad t_2 = \begin{pmatrix} 1 & 2 & 3 \\ 3 & 2 & 1 \end{pmatrix}$$

$$t_3 = \begin{pmatrix} 1 & 2 & 3 \\ 2 & 1 & 3 \end{pmatrix} \quad \sigma_1 = \begin{pmatrix} 1 & 2 & 3 \\ 2 & 3 & 1 \end{pmatrix} \quad \sigma_2 = \begin{pmatrix} 1 & 2 & 3 \\ 3 & 1 & 2 \end{pmatrix}$$

In this notation $\begin{pmatrix} 1 & 2 & 3 \\ 3 & 1 & 2 \end{pmatrix}$ indicates mapping $1 \to 3$, $2 \to 1$, $3 \to 2$.

Compute for example $\sigma_1 t_1$

$$\sigma_1 t_1 = \begin{pmatrix} 1 & 2 & 3 \\ 3 & 2 & 1 \end{pmatrix} = t_2 \in S_3.$$

The image of an element x under permutation σ is denoted by $x\sigma$, i.e. $x\sigma = \sigma(x)$.

The multiplication table for S_3 is

	t_0	t_1	t_2	t_3	σ_1	σ_2
t_0	t_0	t_1	t_2	t_3	σ_1	σ_2
t_1	t_1	t_0	σ_2	σ_1	t_3	t_2
t_2	t_2	σ_1	t_0	σ_2	t_1	t_3
t_3	t_3	σ_2	σ_1	t_0	t_2	t_1
σ_1	σ_1	t_2	t_3	t_1	σ_2	t_0
σ_2	σ_2	t_3	t_1	t_2	t_0	σ_1

Observe that S_3 is not a commutative group because

$$t_1 t_2 \neq t_2 t_1.$$

3.5 THE ALTERNATING GROUPS

DEFINITION OF EVEN AND ODD PERMUTATIONS

$\sigma \in S_n$ is said to be an even permutation if

$$\prod_{i<k} \frac{k\sigma - i\sigma}{k - i} = 1$$

and an odd permutation if

$$\prod_{i<k} \frac{k\sigma - i\sigma}{k - i} = -1$$

EXAMPLE

Consider $\sigma_1 \in S_3$

$$\sigma_1 = \begin{pmatrix} 1 & 2 & 3 \\ 2 & 3 & 1 \end{pmatrix}$$

Then

$$\prod_{i<k} \frac{k\sigma_1 - i\sigma_1}{k - i}$$

$$= \frac{2\sigma_1 - 1\sigma_1}{2 - 1} \cdot \frac{3\sigma_1 - 1\sigma_1}{3 - 1} \cdot \frac{3\sigma_1 - 2\sigma_1}{3 - 2}$$

$$= \frac{3 - 2}{2 - 1} \cdot \frac{1 - 2}{3 - 1} \cdot \frac{1 - 3}{3 - 2} = 1$$

Similarly for $t_2 = \begin{pmatrix} 1 & 2 & 3 \\ 3 & 2 & 1 \end{pmatrix}$ we have

$$\prod_{i<k} \frac{kt_2 - it_2}{k-i}$$

$$= \frac{2t_2 - 1t_2}{2-1} \cdot \frac{3t_2 - 1t_2}{3-1} \cdot \frac{3t_2 - 2t_2}{3-2}$$

$$= \frac{2-3}{2-1} \cdot \frac{1-3}{3-1} \cdot \frac{1-2}{3-2} = -1$$

Hence, σ_1 is even and t_2 is odd.

Each permutation is either even or odd, i.e., for any $\sigma \in S_n$

$$\prod_{i<k} \frac{k\sigma - i\sigma}{k-i} = \pm 1$$

An easy way of determining whether a permutation is even or odd is by counting the number of inversions. In a row of integers the number of integers in the row smaller than the first integer is called the number of inversions. For example, the number of inversions in the row

$$6, 2, 1, 9, 4, 3, 5, 8, 7$$

is 5. Consider the permutation

$$\begin{pmatrix} 1 & 2 & 3 \\ 3 & 2 & 1 \end{pmatrix}$$

The number of inversions in	3, 2, 1	is	2
The number of inversions in	2, 1	is	1
Total number of inversions			3

The permutation is odd.

EXAMPLE

Let
$$\sigma = \begin{pmatrix} 1 & 2 & 3 & 4 & 5 & 6 & 7 \\ 5 & 2 & 1 & 6 & 3 & 7 & 4 \end{pmatrix}$$

We obtain that the number of inversions

in	5, 2, 1, 6, 3, 7, 4	is	4
in	2, 1, 6, 3, 7, 4	is	1
in	1, 6, 3, 7, 4	is	0
in	6, 3, 7, 4	is	2
in	3, 7, 4	is	0
in	7, 4	is	1
Total number of inversions			8

The permutation is even.

The general rule is that if the total number of inversions is even, the permutation is even; and if the total number of inversions is odd, the permutation is odd.

THEOREM 19

The product of

1. Two odd permutations is an even permutation.

2. Two even permutations is even.

3. An even permutation and an odd permutation (or an odd permutation and an even permutation) is odd.

S_n is the set of all permutations of $\{1, 2, ..., n\}$ (or any other set consisting of n elements) which becomes a group when equipped with the operation of multiplication of permutations.

DEFINITION OF ALTERNATING GROUP

The subset of S_n, consisting of all even permutations, is called the alternating group of degree n, and is denoted by A_n.

The set of even permutations A_n forms a subgroup of S_n

1. $A_n \neq 0$ because the identity permutation $\begin{pmatrix} 1 & 2 & ... & n \\ 1 & 2 & ... & n \end{pmatrix} \in A_n$.

2. The operation of multiplication is well-defined, for $\sigma_1 \in A_n$ and $\sigma_2 \in A_n$; $\sigma_1\sigma_2 \in A_n$ (see Theorem 19).

3. If $\sigma_1, \sigma_2 \in A_n$ then

$$\sigma_1\sigma_2^{-1} \in A_n$$

Hence, A_n is a subgroup of S_n.

DEFINITION OF ORDER OF A GROUP

The order of a group, G, is the cardinal number of elements of G. The order of G is denoted by $|G|$ or $o(G)$ or $[G]$.

The order of a group S_n is $n!$

$$|S_n| = n!$$

where $n! = 1 \cdot 2 \cdot 3 \cdot ... \cdot n$.

All permutations in S_n are either odd or even. One can easily show that there are precisely the same number of even permutations as odd ones.

Hence
$$|A_n| = \frac{n!}{2}.$$

THEOREM 20

If n is any positive integer $n > 1$, then
$$|S_n| = n!$$
and
$$|A_n| = \frac{n!}{2}.$$

Note that the set of all odd permutations is not a subgroup of S_n because the identity permutation is an even permutation.

3.6 GROUPS OF ISOMETRIES

Let R be the set of real numbers, and S_R be the symmetric group on R. We shall describe some subgroups of S_R. Since R is the real line, the distance between $a, b \in R$ is defined by
$$d(a, b) = |a - b|.$$

First we shall investigate the group of isometries of the line.

The group of isometries of the line R, denoted by $I(R)$, is the set of all elements of S_R which preserve distance. Hence $f \in I(R)$, if and only if
$$d(af, bf) = d(a, b)$$

for every $a, b \in R$.

$I(R) \neq \phi$ because the identity mapping is an isometry. Suppose $f \in I(R)$, then $f^{-1} \in I(R)$. Indeed $f^{-1} \in S_R$ and

$$d(af^{-1}, bf^{-1}) = d((af^{-1})f, (bf^{-1})f) = d(a, b)$$

If $f, g \in I(R)$, then $g^{-1} \in I(R)$ and

$$d(a(fg^{-1}), b(fg^{-1}))$$
$$= d((af)g^{-1}, (bf)g^{-1})$$
$$= d(af, bf) = d(a, b).$$

Hence, if $f, g \in I(R)$, then $fg^{-1} \in I(R)$. By Theorem 17 we conclude that $I(R)$ is a subgroup of S_R.

$I(R)$ is the group of isometries of R.

THEOREM 21

If $f, g \in I(R)$, and $a, b \in R$, $a \neq b$ exist such that

$$af = ag \quad \text{and} \quad bf = bg$$

then

$$f = g.$$

Now we describe the isometries of R. Suppose $f \in I(R)$, and let

$$0f = a$$

then

$$d(xf, 0f) = d(x, 0) \text{ and}$$

45

Thus
$$|xf - 0f| = |xf - a| = |x|$$

$$xf = \pm x + a.$$

We conclude that if $f \in I(R)$ and $0f = a$, then

$$xf = \pm x + a.$$

Any isometry of the real line is of the form $xf = x + a$ or $xf = -x + a$. Note that $xf = x + a$ moves the real line a units in the positive direction and $xf = -x + a$ inverts the real line, and then moves it a units in the positive direction.

Now we shall discuss the isometries of the plane $R^2 = R \times R$.

The distance between two points $A = (x_A, y_A)$ and $B = (x_B, y_B)$ is defined by

$$d(A, B) = \sqrt{(x_A - x_B)^2 + (y_A - y_B)^2}$$

Let S_{R^2} be the symmetric group of R^2. Similarly, any $f \in S_{R^2}$ is called an isometry if for any $A, B \in R^2$

$$d(A, B) = d(Af, Bf).$$

The following holds:

THEOREM 22

The set $I(R^2)$ of all isometries of R^2 forms a subgroup of S_{R^2}.

There are three basic types of isometries of R^2.

1. A translation $t_{a,b}$ is the mapping defined by

$$(x, y)t_{a,b} = (x + a, y + b).$$

For each a, b the translation $t_{a,b}$ is an isometry and

$$(t_{a,b})^{-1} = t_{-a,-b}.$$

2. A rotation about the origin through an angle α is defined by

$$(x, y)\,\theta_\alpha = (x \cos \alpha - y \sin \alpha,\ x \sin \alpha + y \cos \alpha).$$

θ_α is an isometry for each α and

$$(\theta_\alpha)^{-1} = \theta_{-\alpha}.$$

3. A reflection in the axis OX is defined by

$$(x, y)\,\lambda_{OX} = (x, -y).$$

Also,

$$(\lambda_{OX})^{-1} = \lambda_{OX}.$$

Translation $t_{a,b}$

y ↑ $(x \cos \alpha - y \sin \alpha, x \sin \alpha + y \cos \alpha)$

(x, y)

α

x

Rotation θ_α

y

(x, y)

x

$(x, -y)$

Reflection λ_{ox}

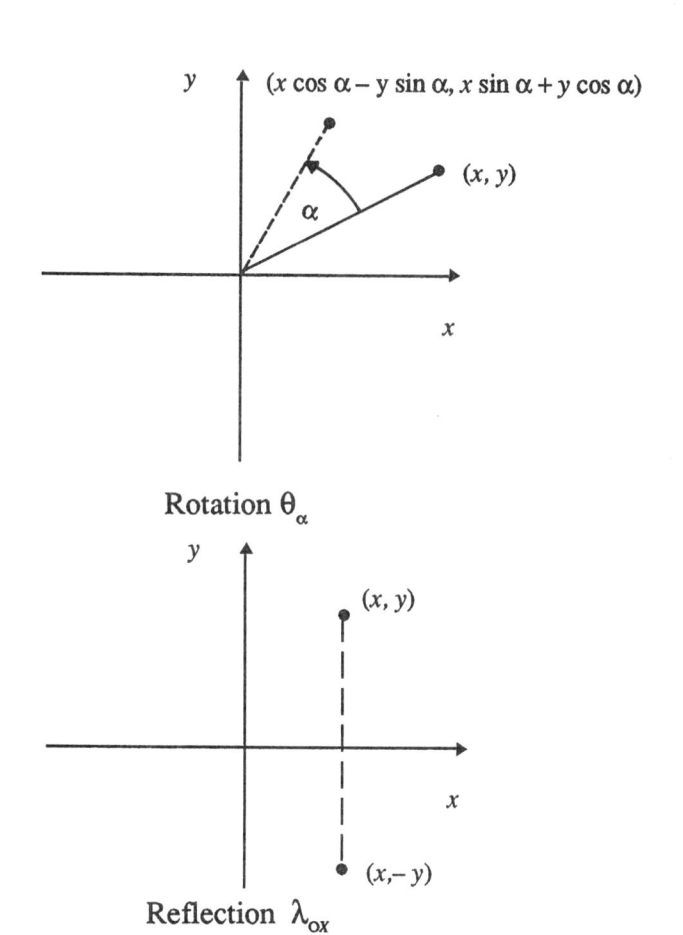

THEOREM 23

Let f and g be two isometries, $f, g \in I(R^2)$, which have the same effect on three noncollinear points in R^2. Then

$$f = g.$$

Note that the three noncollinear points are points which do not lie on a straight line. By Theorem 23 we conclude that an isometry is determined uniquely by its action on any three points which are not collinear.

Finally, we reach the following very important theorem.

THEOREM 24

Every isometry of R^2 can be expressed as the product of a translation, a rotation, and a reflection.

We shall briefly mention the symmetry groups.

The concept of symmetry appears in physics and chemistry. For example, magnetic fields can in some way be symmetric. Also molecular symmetry helps solve the structural problems a chemist might encounter.

The standard approach to all these problems involves defining the symmetry groups.

DEFINITION OF SYMMETRY GROUP

Let P be any subset of the Euclidean plane R^2 and let I_p be the set of all $f \in I(R^2)$ such that

I. If $p \in P$, then $pf \in P$

II. If $tf \in P$, then $t \in P$

Then I_p is a subgroup of $I(R^2)$. I_p is called the symmetry group of P.

EXAMPLE

Find the elements of the symmetry group of the following figure.

In this case $P = ABCD$. I_p contains the following elements:

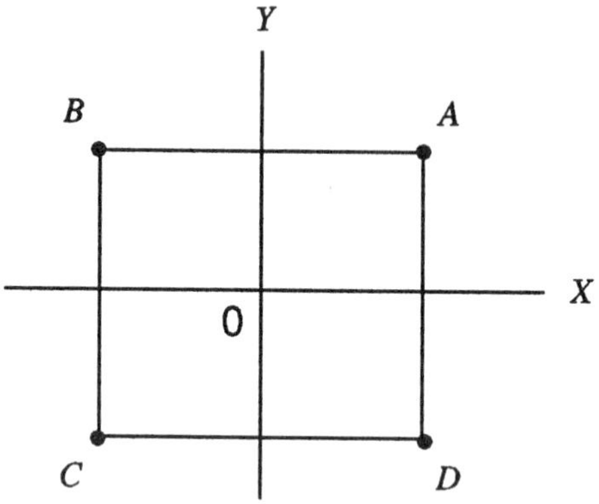

1. i_1 – the identity mapping (rotation about 0 of 0°)

$$Ai_1 = A,\ Bi_1 = B,\ Ci_1 = C,\ Di_1 = D$$

2. i_2 – a clockwise rotation about 0 of 90°

$$Ai_2 = B,\ Bi_2 = C,\ Ci_2 = D,\ Di_2 = A$$

3. i_3 – a clockwise rotation about 0 of 180°

$$Ai_3 = C,\ Bi_3 = D,\ Ci_3 = A,\ Di_3 = B$$

4. i_4 – a clockwise rotation about 0 of 270°

$$Ai_4 = D,\ Bi_4 = A,\ Ci_4 = B,\ Di_4 = C$$

5. i_5 – a reflection in $0X$

$$Ai_5 = D,\ Bi_5 = C$$

6. i_6 – a reflection in $0Y$

$$Ai_6 = B, \ Di_6 = C$$

7. i_7 – a reflection in the diagonal BD

$$Ai_7 = C, \ Bi_7 = B, \ Ci_7 = A$$

8. i_8 – a reflection in the diagonal CA

$$Ai_8 = A, \ Bi_8 = D$$

All these elements of the symmetry group I_{ABCD} are different. It can be shown that I_{ABCD} does not contain any other elements. Thus

$$I_{ABCD} = \{i_1, i_2, i_3, i_4, i_5, i_6, i_7, i_8\} \ .$$

CHAPTER 4

ISOMORPHISMS AND HOMOMORPHISMS

4.1 ISOMORPHISMS

DEFINITION OF ISOMORPHISMS

A one-to-one mapping

$$f: G \to H$$

of G onto H, where both G and H are groups, is called an isomorphism, if whenever

$$g_1 f = h_1$$

and

$$g_2 f = h_2,$$

then

$$g_1 g_2 f = h_1 h_2.$$

EXAMPLE

Let G be the group of integers with the usual addition of integers, and let H be the group of even integers with the usual addition of integers.

The mapping

$$f: G \to H$$

where f is defined by

$$f: g \to 2g$$

is an isomorphism.

For any $h \in H$ there exists a unique $g \in G$ such that $h = 2g$. Hence, f is onto. Also, if $h = 2g_1 = 2g_2$, then $g_1 = g_2$. Hence, f is one-to-one.

Suppose

$$g_1 f = h_1 \quad \text{and} \quad g_2 f = h_2,$$

Then

$$(g_1 + g_2)f = 2g_1 + 2g_2 = g_1 f + g_2 f.$$

Thus f is an isomorphism.

EXAMPLE

$$G_1 \qquad\qquad G_2$$

$$x_1 = \begin{pmatrix} 1 & 2 & 3 \\ 1 & 2 & 3 \end{pmatrix} \qquad y_1 = \begin{pmatrix} 1 & 2 & 3 & 4 & 5 & 6 \\ 1 & 2 & 3 & 4 & 5 & 6 \end{pmatrix}$$

$$x_2 = \begin{pmatrix} 1 & 2 & 3 \\ 2 & 3 & 1 \end{pmatrix} \qquad y_2 = \begin{pmatrix} 1 & 2 & 3 & 4 & 5 & 6 \\ 2 & 3 & 1 & 6 & 4 & 5 \end{pmatrix}$$

$$x_3 = \begin{pmatrix} 1 & 2 & 3 \\ 3 & 1 & 2 \end{pmatrix} \qquad y_1 = \begin{pmatrix} 1 & 2 & 3 & 4 & 5 & 6 \\ 3 & 1 & 2 & 5 & 6 & 4 \end{pmatrix}$$

$$x_4 = \begin{pmatrix} 1 & 2 & 3 \\ 1 & 3 & 2 \end{pmatrix} \quad y_4 = \begin{pmatrix} 1 & 2 & 3 & 4 & 5 & 6 \\ 4 & 5 & 6 & 1 & 2 & 3 \end{pmatrix}$$

$$x_5 = \begin{pmatrix} 1 & 2 & 3 \\ 3 & 2 & 1 \end{pmatrix} \quad y_5 = \begin{pmatrix} 1 & 2 & 3 & 4 & 5 & 6 \\ 5 & 6 & 4 & 3 & 1 & 2 \end{pmatrix}$$

$$x_6 = \begin{pmatrix} 1 & 2 & 3 \\ 2 & 1 & 3 \end{pmatrix} \quad y_6 = \begin{pmatrix} 1 & 2 & 3 & 4 & 5 & 6 \\ 6 & 4 & 5 & 2 & 3 & 1 \end{pmatrix}$$

All permutations of a set form a group. Hence, G_1 is a group. One can show that G_2 forms a subgroup of the full group of permutations. For any $y_i, y_j \in G_2$

$$y_i y_j \in G_2$$

and $y_i^{-1} \in G_2$ where $i, j = 1, 2, \ldots, 6$. For example,

$$y_2 y_4 = y_5.$$

We define

$$f: G_1 \to G_2$$

by $x_i f = y_i$, for $i = 1, 2, \ldots, 6$. Since

$$(x_i x_j) f = y_i y_j$$

f is an isomorphism.

Observe that two groups which are isomorphic are essentially the same.

Let G and H be isomorphic groups. Then an isomorphism exists

54

$$f: G \to H,$$

which is one-to-one and onto by definition and which preserves multiplication.

Each element h of H, $h \in H$ can be renamed by

$$gf = h.$$

We will not distinguish between isomorphic groups. Except for the names of their elements, isomorphic groups are the same.

4.2 HOMOMORPHISMS

DEFINITION OF HOMOMORPHISM

A mapping $f: G \to H$, where G and H are groups, is called a homomorphism, if f is onto and if whenever

$$g_1 f = h_1 \quad \text{and} \quad g_2 f = h_2,$$

then

$$(g_1 g_2)f = h_1 h_2.$$

A one-to-one homomorphism is an isomorphism.

THEOREM 25

Let $f: G \to H$ be a homomorphism.

1. If $e \in G$ is the identity of G, then $ef \in H$ is the identity of H.

2. If $g \in G$ and $gf = h$, then

$$g^{-1}f = h^{-1}$$

Homomorphism maps the identity into identity and inverses into inverses.

EXAMPLE

Let G_1 be the group of permutations

$$x_1 = \begin{pmatrix} 1 & 2 & 3 \\ 1 & 2 & 3 \end{pmatrix} \quad x_4 = \begin{pmatrix} 1 & 2 & 3 \\ 1 & 3 & 2 \end{pmatrix}$$

$$x_2 = \begin{pmatrix} 1 & 2 & 3 \\ 2 & 3 & 1 \end{pmatrix} \quad x_5 = \begin{pmatrix} 1 & 2 & 3 \\ 3 & 2 & 1 \end{pmatrix}$$

$$x_3 = \begin{pmatrix} 1 & 2 & 3 \\ 3 & 1 & 2 \end{pmatrix} \quad x_6 = \begin{pmatrix} 1 & 2 & 3 \\ 2 & 1 & 3 \end{pmatrix}$$

and G_2 be the multiplicative group consisting of 1 and -1. A mapping

$$x_1 \to 1 \quad x_4 \to -1$$
$$x_2 \to 1 \quad x_5 \to -1$$
$$x_3 \to 1 \quad x_6 \to -1$$

is a homomorphism.

THEOREM 26 (Cayley)

Every group G is isomorphic to a permutation group of its own elements.

We shall outline the proof of this theorem.

For each $g \in G$, define the mapping

$$F(g): x \to xg$$

for all $x \in G$. For any fixed g this is a mapping of G onto G. Indeed, let $y \in G$, then $x = yg^{-1} \in G$ and

$$F(g): x \to y$$

$F(g)$ is one-to-one, because if $x_1 g = x_2 g$, then $x_1 = x_2$. Thus $F(g)$ for each $g \in G$ is a permutation of G. The mapping

$$G \ni g \to F(g)$$

is one-to-one and onto. Moreover,

$$F(g_1) F(g_2) = F(g_1 g_2)$$

Thus $g \to F(g)$ is an isomorphism.

Often one-to-one and onto mapping $f: A \to B$ is indicated by

$$f: A \leftrightarrows B$$

We write

$$G \cong H$$

to indicate that two groups G and H are isomorphic.

Homomorphisms and isomorphisms will be discussed later in detail.

4.3 THE SUBGROUP GENERATED BY X

Let X be a subset of a group G. The following notation will be used

$$x_1^{\epsilon_1} \ldots x_k^{\epsilon_k}, \text{ where } \epsilon_i = \pm 1, x_i \in X$$

$$\text{for } i = 1, 2, \ldots, k.$$

EXAMPLE

$x^2 y^{-3} x$ can be written as

$x x y^{-1} y^{-1} y^{-1} x$, thus

$x_1 = x_2 = x$, $\epsilon_1 = \epsilon_2 = 1$

$x_3 = x_4 = x_5 = y$, $\epsilon_3 = \epsilon_4 = \epsilon_5 = -1$

$x_6 = x$, $\epsilon_6 = 1$

EXAMPLE

If
$$a = x_1^{\epsilon_1} \ldots x_n^{\epsilon_n}$$
then
$$a^{-1} = x_n^{-\epsilon_n} \ldots x_1^{-\epsilon_1}$$
Indeed
$$aa^{-1} = x_1^{\epsilon_1} \ldots x_n^{\epsilon_n} x_n^{-\epsilon_n} \ldots x_1^{-\epsilon_1}$$
$$= a^{-1}a = 1$$

H is a subgroup of G if and only if for any $h_1, h_2 \in H$,

$$h_1 h_2^{-1} \in H$$

Let H be a subgroup of G and let $X \subseteq H$. Let $x_1, x_2, \ldots, x_n \in X$ be any elements of X.

Then

$$x_1^{\epsilon_1} \ldots x_n^{\epsilon_n} \in H$$

where n is a positive integer. That leads us to the following.

THEOREM 27

Let G be a group and let $X \neq \phi$ be a subset of G. Then

$$P = \{x_1^{\epsilon_1} \ldots x_n^{\epsilon_n} \mid x_i \in X, e_i = \pm 1, n \text{ a positive integer}\}$$

is a subgroup of G.

Moreover, if H is any subgroup of G and $X \subseteq H$, then

$$P \subseteq H.$$

An arbitrary set of elements in a group is called a complex. If A and B are two complexes in a group G, then

$$AB$$

denotes the complex consisting of all elements ab, $a \in A$, $b \in B$. AB is called the product of A and B.

If X is any complex in a group G, then by $gp(X)$ (or $\{X\}$) we denote the subgroup consisting of all finite products

$$x_1^{\epsilon_1} \ldots x_n^{\epsilon_n},$$

where $x_1, x_2, \ldots, x_n \in X$.

$gp(X)$ is called the subgroup generated by X.

If a group can be generated by a finite set, then a group is called a finitely generated group.

EXAMPLE

Let $(Z, +)$ be the additive group of integers. Then
$$gp(\{1\}) = Z \quad \text{and}$$
$$gp(\{2\}) = \{\ldots, -4, -2\ 0, 2, 4, \ldots\}$$

EXAMPLE

Let $G = \{0, 1, 2\}$ be a group with the multiplication table

	0	1	2
0	0	1	2
1	1	2	0
2	2	0	1

Find $gp(\{1\})$.

Obviously $1 \in gp(\{1\})$.

Since $1 \cdot 1 = 2$ also $2 \in gp(\{1\})$.

Finally $1 \cdot 2 = 0$, hence $0 \in gp(\{1\})$.

Thus
$$gp(\{1\}) = G$$

CHAPTER 5

CYCLIC GROUPS. COSETS

5.1 CYCLIC GROUPS

We write

$$gp(X) = G$$

to indicate that a group G is generated by X.

DEFINITION OF CYCLIC GROUPS

A group which can be generated by a single element is called a cyclic group.

If an element $x \in G$ exists such that

$$gp(x) = G$$

then G is a cyclic group. Each element of a cyclic group $gp(X)$ can be written as

$$x^k$$

where k is any integer. If $a, b \in gp(x)$, then $a = x^k$ and $b = x^l$ and

$$ab = ba.$$

Hence, cyclic groups are abelian.

DEFINITION OF ABELIAN GROUPS

A group, G, such that for all $a, b \in G$

$$ab = ba$$

is called an abelian group.

THEOREM 20

Let G be the cyclic group of order k generated by the element x, $G = gp(x)$. Then

$$G = \{x^0, x^1, \ldots, x^{k-1}\}$$

and $x^k = 1$. Also x^k is the least positive power of x that is 1.

By $x^0 = 1$ we denote the identity of G.

EXAMPLE

In the symmetric group S_n of degree n consider the element

$$\sigma_n = \begin{pmatrix} 1 & 2 & \ldots & n-1 & n \\ 2 & 3 & & n & 1 \end{pmatrix}$$

Then

$$\sigma_n^2 = \begin{pmatrix} 1 & 2 & \ldots & n-2 & n-1 & n \\ 3 & 4 & & n & 1 & 2 \end{pmatrix}$$

$$\vdots$$

$$\sigma_n^n = \begin{pmatrix} 1 & 2 & \ldots & n \\ 1 & 2 & & n \end{pmatrix}$$

Hence all elements $\sigma_n^n, \sigma_n^1, \sigma_n^2, \ldots, \sigma_n^{n-1}$ are distinct and $gp(\sigma_n)$ is cyclic of order n, where n is an arbitrary positive integer.

Thus, for each positive integer $n > 0$, there exists a cyclic group $gp(x)$ of order n.

Suppose G and H are two cyclic groups of the same order. Are they different or similar? The next theorem answers that question.

THEOREM 28

Any two cyclic groups of the same order (finite or infinite) are isomorphic.

We showed that for each $n > 0$ there exists a cyclic group of order n. Let Z be the additive group of integers. Then $gp(1)$ is a cyclic group of infinite order, since

$$gp(1) = Z.$$

Infinite cyclic groups exist, and two infinite groups are isomorphic. We investigate the properties of the cyclic group of order n, or the infinite cyclic group.

Remember that to indicate that two groups are isomorphic we write

$$G \cong H.$$

If x is an element of G, then the order of x is defined as the order of $gp(x)$. Thus, if x is of order $n < \infty$, then $x^n = 1$ and n is the first positive integer for which $x^n = 1$. If x is of infinite order, the $x^n = 1$ implies $n = 0$.

THEOREM 29

If $gp(x)$ is of order $n < \infty$ and

$$x^r = 1,$$

then n divides r.

Let

$$r = an + s$$

where $0 \leq s < n$.

Then

$$1 = x^r = x^{an} x^s = (x^n)^a x^s = x^s = 1$$

and $s = 0$. Thus n divides r.

EXAMPLE

Find the order of

$$t = \begin{pmatrix} 1 & 2 & 3 \\ 2 & 1 & 3 \end{pmatrix} \in S_3$$

and of

$$r = \begin{pmatrix} 1 & 2 & 3 & 4 \\ 2 & 3 & 4 & 1 \end{pmatrix} \in S_4$$

$$t^2 = \begin{pmatrix} 1 & 2 & 3 \\ 2 & 1 & 3 \end{pmatrix}\begin{pmatrix} 1 & 2 & 3 \\ 2 & 1 & 3 \end{pmatrix} = \begin{pmatrix} 1 & 2 & 3 \\ 1 & 2 & 3 \end{pmatrix}$$

Hence the order of t is 2.

$$r^2 = \begin{pmatrix} 1 & 2 & 3 & 4 \\ 2 & 3 & 4 & 1 \end{pmatrix} \begin{pmatrix} 1 & 2 & 3 & 4 \\ 2 & 3 & 4 & 1 \end{pmatrix} = \begin{pmatrix} 1 & 2 & 3 & 4 \\ 3 & 4 & 1 & 2 \end{pmatrix}$$

$$r^3 = \begin{pmatrix} 1 & 2 & 3 & 4 \\ 3 & 4 & 1 & 2 \end{pmatrix} \begin{pmatrix} 1 & 2 & 3 & 4 \\ 2 & 3 & 4 & 1 \end{pmatrix} = \begin{pmatrix} 1 & 2 & 3 & 4 \\ 4 & 1 & 2 & 3 \end{pmatrix}$$

$$r^4 = \begin{pmatrix} 1 & 2 & 3 & 4 \\ 4 & 1 & 2 & 3 \end{pmatrix} \begin{pmatrix} 1 & 2 & 3 & 4 \\ 2 & 3 & 4 & 1 \end{pmatrix} = \begin{pmatrix} 1 & 2 & 3 & 4 \\ 1 & 2 & 3 & 4 \end{pmatrix}$$

The order of r is 4.

EXAMPLE

The group $(Q, +)$ is not cyclic, where Q is the set of rational numbers. Suppose $(Q, +)$ is cyclic. Then there exists $\frac{a}{b} \in Q$, $b \neq 0$ such that

$$gp\left(\frac{a}{b}\right) = Q.$$

We have

$$\frac{1}{2b} \in Q$$

and

$$\frac{1}{2b} = \frac{a}{b} + \ldots + \frac{a}{b} = \frac{ka}{b} \quad \text{or} \tag{1}$$

$$\frac{1}{2b} = -\frac{a}{b} - \ldots - \frac{a}{b} = -\frac{la}{b} \tag{2}$$

where k and l are positive integers.

(1) leads to

$$1 = 2ka$$

which is not true because k and a are integers.

(2) leads to

$$1 = -2la$$

which is not true since l and a are integers. Hence $(Q, +)$ is not cyclic.

5.2 SUBGROUPS OF CYCLIC GROUPS

We shall investigate some basic properties of the subgroups of cyclic groups.

THEOREM 30

For a finite cyclic group, G, there is at most one subgroup of G of any given order.

THEOREM 31

The number of distinct subgroups of the cyclic group, G, is the same as the number of distinct divisors of n, where $n < \infty$ is the order of G.

THEOREM 32

Let H be a subgroup of a cyclic group $G = gp(x)$. Then subgroup H is cyclic and two possibilities exist.

1. $H = \{1\}$ or

2. $H = gp(x^l)$ where x^l is the least positive power of x such that $x^l \in H$.

Let $n < \infty$ be the order of G. Then $l \mid n$ and the order of H is n/l. If the order of G is infinite, then the order of H is infinite or $H = \{1\}$.

THEOREM 33

If l is any positive integer dividing n, then $A = gp(x^l)$ is of order n/l. Thus, there is a subgroup of order k for any k that divides n.

The above theorems establish a fairly thorough knowledge of cyclic groups and their subgroups. For the finite cyclic groups we can determine distinct subgroups which are cyclic. Infinite cyclic groups have infinitely many subgroups.

EXAMPLE

If G is a cyclic group of order a prime p, then G has two subgroups only, $\{1\}$ and G, i.e. G has no proper subgroups. Indeed, the number of subgroups of G is the same as the number of distinct divisors of p. Since p is a prime the only divisors are 1 and p.

Hence G has two subgroups, $\{1\}$ and G.

In the case of finite cyclic groups we can easily determine the number of subgroups and their order. We also know that all subgroups of a cyclic group are cyclic.

Next we shall introduce the important concept of a coset.

5.3 COSETS

DEFINITION OF COSETS

Let G be a group and H its subgroup. A right coset of H in G

is a subset of the form

$$Hg = \{x \mid x = hg, h \in H, g \in G, g \text{ fixed}\}.$$

Similarly, a left coset is defined as

$$gH = \{x \mid x = gh, h \in H, g \text{ fixed}, g \in G\}.$$

EXAMPLE

Let G be the cyclic group of order 4 generated by $\{a\}$ and let $H = gp(a^2)$.

The right cosets of H in G are

$$H1 = H = \{1, a^2\}$$

$$Ha = \{a, a^3\}$$

$$Ha^2 = \{a^2, a^4\} = \{1, a^2\} = H$$

Remember that $a^4 = 1$.

Similarly, we can show that $Ha^3 = Ha$. Hence the distinct cosets of H in G are H and Ha.

$$H \cap Ha = \phi.$$

The union of cosets is G

$$H \cup Ha = G = \{1, a, a^2, a^3\}.$$

THEOREM 34

Two left (or right) cosets of H in G are either disjoint or identical sets of elements. A left (or right) coset of H contains the same cardinal number of elements as H.

The element $g = g1 = 1g$ belongs to the cosets gH and Hg and is called the representative of the coset.

The conclusion from the above theorem is that:

The right (left) cosets of H in G form a partition of G. That is, the union of all the right (left) cosets of H in G is G itself. The intersection of any pair of distinct cosets is empty. We write

$$G = H + Ha_2 + \ldots + Ha_r$$

to indicate that the cosets $H, Ha_2, \ldots, Ha_r$ form a partition of G.

DEFINITION OF INDEX

The cardinal number, r, of right (or left) cosets of a subgroup H in a group, G, is called the index of H in G and is written $[G : H]$.

THEOREM 35 (Theorem of Lagrange)

The order of a group, G, is the product of the order of a subgroup, H, and the index of H in G.

$$|G| = |H| \cdot [G : H].$$

The Theorem of Lagrange plays an important role in the theory of groups. It has many important consequences. We shall list some of them in the form of theorems.

THEOREM 36

Let H be a subgroup of G, and K a subgroup of H, $K \subseteq H \subseteq G$, then

$$[G:K] = [G:H] \cdot [H:K].$$

Since H is a subgroup of G

$$|G| = |H|[G:H].$$

Since K is a subgroup of H

$$|H| = |K|[H:K].$$

Also K is a subgroup of G

$$|G| = |K|[G:K].$$

Hence

$$[G:K] = \frac{|G|}{|K|} = \frac{|H|[G:H]}{|K|} = \frac{|K|[H:K][G:H]}{|K|}$$
$$= [G:H][H:K]$$

Since K is a subgroup $1 \in K$ and $|K| \neq 0$.

THEOREM 37

Let G be a group not the identity alone. Then G has no subgroup except itself and the identity if, and only if, G is a finite cyclic group of prime order.

EXAMPLE

Show that S_7 has no subgroup of order 15. We have

$$|S_7| = 7! = 2 \cdot 3 \cdot 4 \cdot 5 \cdot 6 \cdot 7 = 2^4 \cdot 3^2 \cdot 5 \cdot 7.$$

If H were a subgroup of S_7 of order 13, then 13 divides 7! but 13 does not divide $2^4 \cdot 3^2 \cdot 5 \cdot 7$.

EXAMPLE

If G is a group of prime order, then G is cyclic.

5.4 NORMAL SUBGROUPS

THEOREM 38

A necessary and sufficient condition for the right cosets of H in G to yield the same partition as the left cosets is that for each $a \in G$

$$aH = Ha$$

DEFINITION OF NORMAL SUBGROUPS

A subgroup, H, of a group, G, is said to be a normal (or invariant) subgroup of G if and only if

$$aH = Ha$$

for every element $a \in G$.

Note that $aH = Ha$ for all $a \in G$ if and only if $a^{-1}ha \in H$ for all $h \in H$ and for all $a \in G$. Hence another definition of normal subgroups is:

A subgroup, H, of a group, G, is normal in G if $g^{-1}hg \in H$ for all $h \in H$ and all $g \in G$.

EXAMPLE

Every subgroup of an abelian group is a normal subgroup.

Let G be an abelian group and H its subgroup. Since G is abelian, for any $g \in G$ and $h \in H$

$$h = g^{-1} h g$$

because $gh = hg$.

If $h \in H$, then $g^{-1}hg \in H$ and H is a normal subgroup of G.

EXAMPLE

A subgroup of order 2 is necessarily a normal subgroup.

We shall define some typical normal subgroups.

DEFINITION OF THE CENTER OF G

If G is a group, the center of G, denoted by $Z(G)$ is defined by

$$Z(G) = \{x \mid x \in G \text{ and for all } g \in G, gx = xg\}$$

Observe that $Z(G)$ is a normal subgroup of G.

For all $g \in G$, $1g = g1$. Hence $1 \in Z(G)$. If $g_1, g_2 \in Z(G)$ then $g_1 g_2^{-1} \in Z(G)$.

Hence $Z(G)$ is a subgroup of G. If $g \in G$ and $z \in Z(G)$, then $gz = zg$ and $z = g^{-1} zg$.

Thus $g^{-1} zg \in Z(G)$ for all $g \in G$ and for all $z \in Z(G)$. Hence $Z(G)$ is a normal subgroup of G.

DEFINITION OF CENTRALIZER

The centralizer $C(A)$ of A in G is defined by

$$C(A) = \{y \mid y \in G \text{ and for all } a \in A, ya = ay\}.$$

It can be shown that the centralizer $C(A)$ is a subgroup of G. Also, if A is an abelian subgroup of G, then A is normal in $C(A)$.

DEFINITION OF NORMALIZER

The normalizer $N(A)$ of A in G is defined by

$$N(A) = \{x \mid x \in G \text{ and } xA = Ax\}.$$

$N(A)$ is a subgroup of G.

If A is a subgroup of G, then A is normal in $N(A)$.

We shall end this chapter with some examples of groups.

5.5 EXAMPLES OF GROUPS

DIHEDRAL GROUPS

The symmetries of a regular polygon of $n \geq 3$ sides form a group of order $2n$. Number the vertices clockwise $1, 2, \ldots, n$. All symmetries are generated by the rotation

$$w = \begin{pmatrix} 1 & 2 & 3 & n-1 & n \\ 2 & 3 & 4 & n & 1 \end{pmatrix}$$

and the reflection

$$r = \begin{pmatrix} 1 & 2 & 3 & n-1 & n \\ 1 & n & n-1 & 3 & 2 \end{pmatrix}$$

Here

$$w^n = 1, r^2 = 1$$

$$rw = w^{-1} r$$

and these relations determine the group of order $2n$ completely.

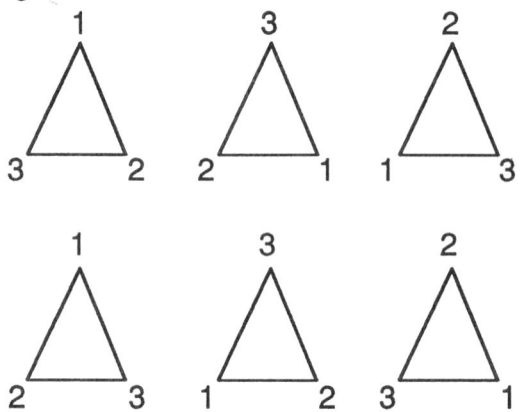

For a triangle

SYMMETRIES OF THE CUBE

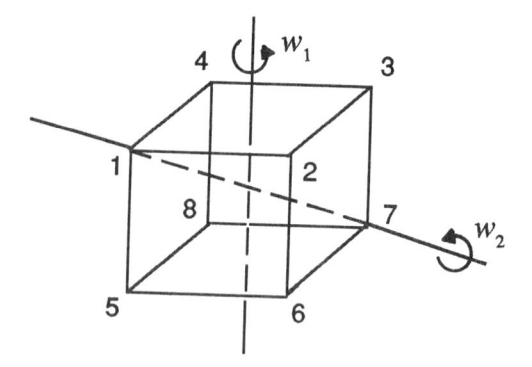

All the symmetries are determined by the mapping of the eight vertices onto themselves. Consider the rotations

$$w_1 = \begin{pmatrix} 1 & 2 & 3 & 4 & 5 & 6 & 7 & 8 \\ 2 & 3 & 4 & 1 & 6 & 7 & 8 & 5 \end{pmatrix}$$

$$w_2 = \begin{pmatrix} 1 & 2 & 3 & 4 & 5 & 6 & 7 & 8 \\ 1 & 4 & 8 & 5 & 2 & 3 & 7 & 6 \end{pmatrix}$$

and the reflection

$$r = \begin{pmatrix} 1 & 2 & 3 & 4 & 5 & 6 & 7 & 8 \\ 5 & 6 & 7 & 8 & 1 & 2 & 3 & 4 \end{pmatrix}$$

The elements w_1 and w_2 generate a group G_0 which takes every vertix into every other vertex. By H_0 we denote a subgroup whose elements keep 1 fixed. The index of H_0 in G_0 is 8, $[G_0 : H_0] = 8$.

The order of H_0 is 3. Hence

$$|G_0| = |H_0| \cdot [G_0 : H_0] = 3 \cdot 8 = 24.$$

The reflection r is not in G_0

$$G = G_0 + G_0 r$$

Thus the order of the full group of symmetries of the cube is 48.

THE QUARTERNION GROUP

The group of order 8 with the following multiplication table:

	a_1	a_2	a_3	a_4	a_5	a_6	a_7	a_8
a_1	a_1	a_2	a_3	a_4	a_5	a_6	a_7	a_8
a_2	a_2	a_5	a_4	a_7	a_6	a_1	a_8	a_3
a_3	a_3	a_8	a_5	a_2	a_7	a_4	a_1	a_6
a_4	a_4	a_3	a_6	a_5	a_8	a_7	a_2	a_1
a_5	a_5	a_6	a_7	a_8	a_1	a_2	a_3	a_4
a_6	a_6	a_1	a_8	a_3	a_2	a_5	a_4	a_7
a_7	a_7	a_4	a_1	a_6	a_3	a_8	a_5	a_2
a_8	a_8	a_7	a_2	a_1	a_4	a_3	a_6	a_5

In a product $xy = z$ any two of a, b, c determine the third uniquely.

Hence, the table defines a quasi-group.

DEFINITION OF QUASI-GROUP

A quasi-group, A, is a system of elements $A(a, b, c, \ldots)$ in which a binary operation is defined in such a way that in

$$ab = c$$

any two of a, b, c determine the third uniquely, $a, b, c \in A$.

Also for every a_k

$$a_k a_1 = a_1 a_k = a_k$$

Hence, the table determines a loop.

DEFINITION OF LOOP

A quasi-group with a unit 1 such that

$$1a = a1 = a$$

for every element a is called a loop.

Note that to prove the associative law we should check $8 \cdot 8 \cdot 8 = 512$ equations

$$a_l(a_m a_n) = (a_l a_m) a_n.$$

CHAPTER 6

HOMOMORPHISMS

6.1 THE KERNEL OF A HOMOMORPHISM

Suppose the group, H, is a homomorphic image of the group, G, $f: G \to H$. The kernel of the homomorphism, denoted by ker f is the set of elements of G, $t \in G$, consisting of all elements of G mapped onto the identity of H

$$\ker f = \{t \mid t \in G, f(t) = 1\}.$$

If $f: G \to H$ is a homomorphism, then

$$f: 1 \to 1,$$

hence $1 \in \ker f$.

If $f: g \to 1$ and $g^{-1} \to h$, then

$$1 = gg^{-1} \to 1 \cdot h = h.$$

Hence $h = 1$ and $g^{-1} \in \ker f$ if $g \in \ker f$.

If $g_1 \to 1$, and $g_2 \to 1$, then

Thus
$$g_1 g_2 \to 1.$$

$$g_1 g_2 \in \ker f.$$

We conclude that the kernel of a homomorphism $f: G \to H$ is a subgroup of G.

If $g \in G$ and $t \in \ker f$, then

$$g \to h, \quad t \to 1, \quad g^{-1} \to h^{-1},$$

and

$$g^{-1} t g \to h^{-1} 1 h = 1,$$

hence,

$$g^{-1} t g \in \ker f.$$

The kernel is a normal subgroup of G.

THEOREM 39 (First Theorem on Homomorphisms)

The kernel of a homomorphism

$$f: G \to H$$

is a normal subgroup of G. Two elements of G have the same image in H if and only if they belong to the same coset of $\ker f$.

6.2 FACTOR GROUPS

Let T be the kernel of a homomorphism $f: G \to H$. Then T is a normal subgroup of G. Conversely, every normal subgroup T of

G is the kernel of a homomorphism, which is unique.

Suppose
$$G = T + Tx_2 + \ldots + Tx_n$$
where T is a normal subgroup of G.

We construct a new group, H, whose elements are cosets Tx_i. The product is defined by
$$(Tx_i)(Tx_j) = Tx_k$$
if $x_i x_j \in Tx_k$ in G.

It can be shown that the product is uniquely defined.

Since T is a normal subgroup
$$Tx_i = x_i T$$
$$T^2 = T.$$
Hence in H, T is a unit since
$$T \cdot Tx_i = Tx_i \quad \text{and}$$
$$Tx_i \cdot T = TTx_i = Tx_i.$$

The product is associative.

Also if
$$x_i^{-1} \in Tx_j \quad \text{then}$$
$$1 = x_i x_i^{-1} \in Tx_i Tx_j.$$

Hence $Tx_i Tx_j = T$ and Tx_j is the inverse of Tx_i.

DEFINITION OF FACTOR GROUP

Group H, described above, is called the factor group of G with respect to T and is denoted by

$$H = \frac{G}{T}$$

THEOREM 40 (Second Theorem on Homomorphisms)

Let G be a group and T a normal subgroup. If

$$H = \frac{G}{T}$$

then there is a homomorphism

$$f: G \to H$$

(called the natural homomorphism of G onto its factor group $\frac{G}{T}$) whose kernel is T.

This homomorphism is defined by

$$f: g \to Tx_i \quad \text{if} \quad g \in Tx_i \text{ in } G.$$

THEOREM 41 (Third Theorem on Homomorphisms)

If $f: G \to K$ is a homomorphism and T is the kernel of the homomorphism, then K is isomorphic to $H = \frac{G}{T}$.

If $f: g \to g'$, then $h: g' \to Tg$ is an isomorphism between K and H.

THEOREM 42

If A and B are subgroups of a group G, and either one of them is a normal subgroup, then

$$A \cup B = AB.$$

CHAPTER 7

THE SYLOW THEOREMS

7.1 MORE ABOUT THE THEOREM OF LAGRANGE

According to the Theorem of Lagrange, the order of a subgroup of a finite group divides the order of a group. Suppose G is a group of finite order and $k \mid \mid G \mid$ We shall find the answer to the question:

Is there always a subgroup of order k?

The following examples show that the answer to that question is no.

Consider the permutation group of order 12.

$$\begin{pmatrix} 1 & 2 & 3 & 4 \\ 1 & 2 & 3 & 4 \end{pmatrix} \begin{pmatrix} 1 & 2 & 3 & 4 \\ 1 & 3 & 4 & 2 \end{pmatrix} \begin{pmatrix} 1 & 2 & 3 & 4 \\ 2 & 1 & 4 & 3 \end{pmatrix} \begin{pmatrix} 1 & 2 & 3 & 4 \\ 1 & 4 & 2 & 3 \end{pmatrix}$$

$$\begin{pmatrix} 1 & 2 & 3 & 4 \\ 3 & 4 & 1 & 2 \end{pmatrix} \begin{pmatrix} 1 & 2 & 3 & 4 \\ 3 & 2 & 4 & 1 \end{pmatrix} \begin{pmatrix} 1 & 2 & 3 & 4 \\ 4 & 3 & 2 & 1 \end{pmatrix} \begin{pmatrix} 1 & 2 & 3 & 4 \\ 4 & 2 & 1 & 3 \end{pmatrix}$$

$$\begin{pmatrix} 1 & 2 & 3 & 4 \\ 2 & 3 & 1 & 4 \end{pmatrix} \begin{pmatrix} 1 & 2 & 3 & 4 \\ 2 & 4 & 3 & 1 \end{pmatrix} \begin{pmatrix} 1 & 2 & 3 & 4 \\ 3 & 1 & 2 & 4 \end{pmatrix} \begin{pmatrix} 1 & 2 & 3 & 4 \\ 4 & 1 & 3 & 2 \end{pmatrix}$$

The divisors of the order of the group 12 are 2, 3, 4, and 6. One can verify that the subgroups of order 2, 3, and 4 exist but there is no subgroup of order 6.

In general, if m divides n, m/n, we cannot be sure that a group of order n contains a subgroup of order m. A question arises. How does one determine the existence and number of subgroups of a group of finite order?

This question is answered by the Sylow Theorem. Throughout this book p denotes a prime number.

7.2 THE SYLOW THEOREM

THEOREM 43

If the order of a group, G, is divisible by a prime, p, then G contains an element of order p.

This theorem guarantees the existence of at least one subgroup of order p whenever p divides the order of G.

THEOREM 44 (First Sylow Theorem)

If G is of order n

$$n = p^m s$$

where p is a prime and

$$p \mid s$$

(p does not divide s), then G contains subgroups of orders

$$p^i, i = 1, 2, \ldots m.$$

Let H be a subgroup of G of order a power of a prime, p, and let $|H|$ be the highest power of p that divides $|G|$. H is called a Sylow p-subgroup of G.

Every finite group has a Sylow p-subgroup.

DEFINITION OF P-GROUPS

A group of order a power of the prime p is called a p-group.

EXAMPLE

Consider the symmetric group S_3, which consists of six elements.

$$t_0 = \begin{pmatrix} 1 & 2 & 3 \\ 1 & 2 & 3 \end{pmatrix} \quad t_1 = \begin{pmatrix} 1 & 2 & 3 \\ 1 & 3 & 2 \end{pmatrix} \quad t_2 = \begin{pmatrix} 1 & 2 & 3 \\ 3 & 2 & 1 \end{pmatrix}$$

$$t_3 = \begin{pmatrix} 1 & 2 & 3 \\ 2 & 1 & 3 \end{pmatrix} \quad S_1 = \begin{pmatrix} 1 & 2 & 3 \\ 2 & 3 & 1 \end{pmatrix} \quad S_2 = \begin{pmatrix} 1 & 2 & 3 \\ 3 & 1 & 2 \end{pmatrix}$$

$$|S_3| = 6.$$

Since $|S_3| = 6 = 2 \cdot 3$ the order of any Sylow 2-subgroup is 2 and the order of any Sylow 3-subgroup is 3. Since $t_1^2 = t_2^2 = t_3^2 = t_0$ the subgroups of order two are

$$\{t_0, t_1\}, \{t_0, t_2\}, \{t_0, t_3\}.$$

By definition they are Sylow-2 subgroups of S_3.

$\{t_0, S_1, S_2\}$ is the only subgroup of order three in S_3, thus it is the only Sylow-3 subgroup of S_3.

THEOREM 45 (Second Sylow Theorem)

If H is a subgroup of a finite group, G, and H is a p-group, then H is contained in a Sylow-p subgroup of G.

For the next theorem we need the following.

DEFINITION OF CONJUGATE SUBGROUPS

Two subgroups A and B of a group G are called conjugate if there is a $g \in G$ such that

$$g^{-1} A g = B.$$

THEOREM 46 (Third Sylow Theorem)

In a finite group, G, the Sylow p-subgroups are conjugate. The number of Sylow p-subgroups of a finite group, G, is of the form $1 + kp$ and is a divisor of the order of G.

The Sylow theorems have numerous applications.

EXAMPLE

There is one and only one group of order 15. Suppose $|G| = 15$, then G has at least one subgroup of order 3 and at least one of order 5.

From Theorem 46 we conclude that there are $1 + 3k$ subgroups of order 3 and $1 + 3k \mid |G|$, i.e. $1 + 3k \mid 15$. That implies $k = 0$. Hence G has $1 + 3 \cdot 0 = 1$ one and only one subgroup of order 3. Similarly, G has one and only one subgroup of order 5.

These subgroups must be cyclic

$$A = \{1, a, a^2\}$$
$$B = \{1, b, b^2, b^3, b^4\}$$

The order of ab in G must be either 1, 3, 5, or 15. One can show that the order of ab is 15. Thus G is the cyclic group generated by ab of order 15.

EXAMPLE

The alternating group, A_4, has no subgroup of order 6.

We shall introduce two important concepts related to the Sylow Theorems.

DEFINITION OF NORMALIZER

Let A be a non-empty subset of a group G. The normalizer of A in H, denoted $N_H(A)$, is the set

$$N_H(A) = \{h \mid h \in H, h^{-1}Ah = A\}$$

where H is a subgroup of G. The normalizer $N_H(A)$ is a subgroup of H.

If $G = H$ we have the normalizer, $N(A)$ of A in a group G, as defined by

$$N(A) = \{g \mid g \in G, Ag = gA\},$$

which is the definition given in 5.4.

DEFINITION OF H-CONJUGATE

Let A and B be non-empty subsets of a group, G, and let H be a subgroup of G. B is said to be an H-conjugate of A if

$$h^{-1}Ah = B$$

for some $h \in H$.

If $G = H$ then A and B become conjugate as defined previously.

Now suppose a group, G, and its subgroup, H, are given. Let A be a non-empty subset of G. What is the number of distinct subsets of G which are H-conjugates of A?

THEOREM 47

If G is a finite group with subgroup H and non-empty subset A, then the number of distinct H-conjugates of A is the index of $N_H(A)$ in H, that is $[H : N_H(A)]$.

7.3 THE CLASS EQUATION OF G

Let Ω be a set of subsets of G. For $A, B \in \Omega$ we define a relation $A \sim B$ by

$$A \sim B, \text{ if } B \text{ is an } H\text{-conjugate of } A$$

i.e. if there exists an element $h \in H$ such that

$$h^{-1}Ah = B \ (H \text{ is a subgroup of } G).$$

The relation $\sim$ is an equivalence relation on Ω.

Thus, $\sim$ defines the partition of Ω into equivalence classes. Note that

$$[A] = \{B \mid B \sim A, A \in \Omega, B \in \Omega\}$$

is the equivalence class containing A.

DEFINITION OF A SET OF REPRESENTATIVES

A set of representatives of the equivalence classes is a set, S, which contains one, and only one, element from each of the distinct equivalence classes.

Hence we can write:

$$o(\Omega) = \sum_{A \in S} o([A])$$

THEOREM 48

Let $\Omega \neq \phi$ be a set of subsets of G and $\sim$ be the equivalence relation on Ω defined by $A \sim B$ if B is an H-conjugate of A. We assume that for each $A \in \Omega$ and each $h \in H$ (H is a subgroup of G)

$$h^{-1}Ah \in \Omega.$$

Then

$$o(\Omega) = \sum_{s \in S} [H : N_H(S)]$$

where S is a set of representatives of the equivalence classes.

Suppose $A \neq \phi$ is a subset of G and

$$\Omega = \{g^{-1}Ag \mid g \in G\}$$

then

$$o(\Omega) = \sum_{s \in S} [H : N_H(S)] = [G : N_G(A)]$$

THEOREM 49

Let Ω be a set consisting of subsets of G which contain one, and only one element, and let ~ be the equivalence relation in Ω when $H = G$, and let S be the set of representatives of the equivalence classes. Define

$$T = \{s \mid s \in S, s \cap Z(G) = \phi\}$$

where

$$Z(G) = \{x \mid xg = gx \text{ for all } g \in G\}.$$

Then

$$o(G) = o(Z(G)) + \sum_{s \in T} [G : N_G(S)]$$

Let $C(s)$ be the centralizer of s in G then

$$o(G) = o(Z(G)) + \sum_{s \in T} [G : C(S)]$$

This equation is called the class equation of G.

CHAPTER 8

FINITE p-GROUPS

8.1 CONSTRUCTION OF FINITE p-GROUPS

It follows from the Sylow theorems that a group G of order

$$n = p_1^{n_1} \cdot \ldots \cdot p_r^{n_r}$$

contains for each $k = 1, \ldots, r$ a subgroup of order $p_k^{n_k}$. All subgroups of this order are isomorphic and conjugate.

The structure of the Sylow p-subgroups of G partly determines the structure of G.

The problem of constructing finite groups consists of two parts:

1. Constructing groups of prime power order;

2. Combining groups of prime power orders dividing a number n to form a group of order n.

When all the Sylow p-subgroups are cyclic, the second problem can be solved.

THEOREM 50

If G is a finite group whose Sylow p-subgroups are all cyclic, then G has a normal subgroup N such that G/N and N are both cyclic.

We shall list some important properties of p-groups (groups of order a power of p, where p is a prime number).

THEOREM 51

The center of a finite p-group is greater than the identity alone. That is

$$Z(G) \neq \{1\}.$$

Also we have: if G is a group of order p^r, $r \geq 1$, then G has a normal subgroup of order p^{r-1}.

THEOREM 52

Every proper subgroup of a p-group G of order p^m is contained in a maximal subgroup of order p^{m-1}, and all the maximal subgroups of G are normal subgroups.

One of the consequences of the Sylow theorems is that no proper subgroup of a p-group is its own normalizer. The following is a converse.

THEOREM 53

In a finite group G the property that no proper subgroup is its own normalizer is true if, and only if, G is the direct product of its Sylow subgroups.

THEOREM 54

If A is a normal subgroup of order p contained in the p-group G, then A is in the center of G.

EXAMPLE

Let G be the additive group of integers modulo p

$$G = \{0, 1, ..., p-1\},$$

and let H be the additive group of integers modulo p^2

$$H = \{0, 1, ..., p^2 - 1\}.$$

We define

$$K = \{(i, j) \mid i \in G, j \in H\}.$$

Under the binary operation

$$(i, j) \cdot (i_1, j_1) = (i + i_1, \; j + j_1 + ji_1 \, p).$$

K is a non-abelian group of order p^3. Obviously, K is of order p^3. Also, K is a semigroup.

$$((i, j) \cdot (i_1, j_1)) \cdot (i_2, j_2) = (i, j) \cdot ((i_1, j_1) \cdot (i_2, j_2)).$$

The binary operation in G is associative.

The identity element of K is

$$(0, 0).$$

The inverse of (i, j) is

$$(-i, -j + jip).$$

Hence K is a group. K is non-abelian because

$$(1, 0)(0, 1) = (1, 1)$$
$$(0, 1)(1, 0) = (1, 1+p).$$

8.2 GROUPS OF ORDER p, p^2, pq, p^3

Suppose G and H are groups and $G \times H$ is the Cartesian product. Let $(g_1, h_1), (g_2, h_2) \in G \times H$; we define the product as

$$(g_1, h_1) \cdot (g_2, h_2) = (g_1 g_2, h_1 h_2). \qquad (*)$$

DEFINITION OF DIRECT PRODUCT

The group $G \times H$; (where G and H are groups) with binary operation defined by (*), is called the external direct product of the groups G and H.

If both G and H are finite groups then

$$|G \times H| = |G| \cdot |H|.$$

The direct product yields a method of constructing new groups.

EXAMPLE

Let C_2 be the cyclic group of order 2 generated by g. Then

$$C_2 \times C_2 = \{(1,g), (g,g), (g,1), (1,1)\}$$

and $|C_2| = 2$, $|C_2 \times C_2| = 4$.

There are two groups of order 4: the cyclic group of order 4

$C_4 = \{1, a, a^2, a^3\}$ where $a^4 = 1$

and the group $C_2 \times C_2$ with the multiplication table

	(1,1)	(1,g)	(g,1)	(g,g)
(1,1)	(1,1)	(1,g)	(g,1)	(g,g)
(1,g)	(1,g)	(1,1)	(g,g)	(g,1)
(g,1)	(g,1)	(g,g)	(1,1)	(1,g)
(g,g)	(g,g)	(g,1)	(1,g)	(1,1)

All elements of $C_2 \times C_2$ are of order 2, thus C_4 and $C_2 \times C_2$ are not isomorphic. $C_2 \times C_2$ is called the Klein four group.

THEOREM 55

Let $G = H \times K$ be the direct product of the groups H and K, and let

$$\overline{H} = \{(h, 1) \mid h \in H, \text{1 the identity of } H\}$$

$$\overline{K} = \{(1, k) \mid k \in K, \text{1 the identity of } K\}.$$

Then $\overline{H}$ and $\overline{K}$ are subgroups of G and

$$H \cong \overline{H} \quad (H \text{ and } \overline{H} \text{ are isomorphic})$$

$$K \cong \overline{K} \quad (K \text{ and } \overline{K} \text{ are isomorphic}).$$

Furthermore, if $\overline{h} \in \overline{H}$ and $\overline{k} \in \overline{K}$, then

$$\overline{h}\,\overline{k} = \overline{k}\,\overline{h}.$$

Also $G = \overline{H}\,\overline{K}$ and $\overline{H} \cap \overline{K}$ $\{(1, 1)\}$.

Observe that for every $g \in G$, g can be uniquely written as a product

$$g = \overline{h}\,\overline{k} \text{ where } \overline{h} \in \overline{H}, \overline{k} \in \overline{K}.$$

The following holds:

THEOREM 56

Let G be a group with subgroups H and K such that

$$H \cap K = \{1\},$$

the elements of H commute with the elements of K and $G = HK$. Then

$$G \cong H \times K.$$

We can write Theorem 56 in the form: Let G be a group with normal subgroups H and K, such that

$$H \cap K = \{1\} \text{ and } G = HK.$$

Then

$$G \cong H \times K.$$

Observe that if G is a finite group with subgroups H and K, then

$$|H|\cdot|K| = |H \cap K|\cdot|HK|.$$

95

EXAMPLE

Suppose G is a group of order 21 and H and K are subgroups of G of orders 3 and 7, respectively.

An element which belongs to $H \cap K$ must have order dividing 7 and 3.

Thus
$$H \cap K = \{1\}$$
and
$$|HK| = \frac{|H| \cdot |K|}{|H \cap K|} = |H| \cdot |K| = |G| = 21.$$

For the finite groups:

THEOREM 57

Let G be a finite group with normal subgroups H and K where $|H| \cdot |K| = |G|$. If either

1. $H \cap K = \{1\}$ or

2. $HK = G$, then

$$G \cong H \times K.$$

Here is one important property of the subgroups of a group.

THEOREM 58

If H and K are subgroups of a group G such that

(*) $hk = kh$ for all $h \in H$ and $k \in K$

(**) every element $g \in G$ is a unique product $g = hk$ of an element $H \ni h$ and an element in $K \ni k$

then

$$G \cong H \times K.$$

DEFINITION OF INTERNAL DIRECT PRODUCT

G is said to be the internal direct product of H and K, written $G = H \otimes K$ if G is a group and H and K are subgroups of G satisfying conditions (*) and (**).

By Theorem 58

$$H \otimes K \cong H \times K.$$

EXAMPLES

I. G order p

A group of prime order p cannot have proper subgroups, thus must be a cyclic group generated by any element different from the identity.

G cyclic, $a^p = 1$.

II. G order p^2

If it is not cyclic, it will contain two distinct subgroups of order p, say $\{a\}$ and $\{b\}$, such that $a^p = 1$, $b^p = 1$. G is an abelian group with a, b as a basis. Hence:

1. G is cyclic, $a^{p^2} = 1$

2. Elementary abelian, $a^p = 1$, $b^p = 1$, $ab = ba$.

III. G order pq, where $p < q$ are primes.

1. G is cyclic, $a^{pq} = 1$.

2. Non-abelian, $a^p = 1$, $b^q = 1$, $a^{-1}ba = b^r$ and $r^p \equiv 1 \pmod{q}$, $r \not\equiv 1 \pmod{q}$, p divides $q - 1$, $p \mid q - 1$.

IV G order p^3

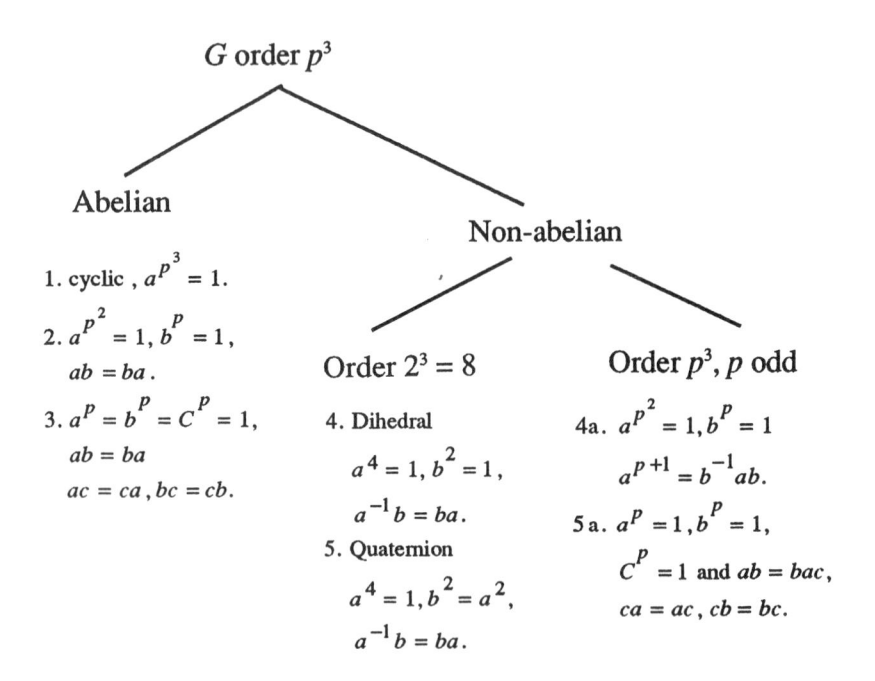

G order p^3

Abelian

Non-abelian

1. cyclic, $a^{p^3} = 1$.

2. $a^{p^2} = 1$, $b^p = 1$, $ab = ba$.

3. $a^p = b^p = c^p = 1$, $ab = ba$, $ac = ca$, $bc = cb$.

Order $2^3 = 8$

4. Dihedral
$a^4 = 1$, $b^2 = 1$,
$a^{-1}b = ba$.

5. Quaternion
$a^4 = 1$, $b^2 = a^2$,
$a^{-1}b = ba$.

Order p^3, p odd

4a. $a^{p^2} = 1$, $b^p = 1$
$a^{p+1} = b^{-1}ab$.

5a. $a^p = 1$, $b^p = 1$,
$c^p = 1$ and $ab = bac$,
$ca = ac$, $cb = bc$.

From the chart, we can determine the number of distinct groups of each order.

For example, there are five non-isomorphic groups of order 8, three abelian and two non-abelian.

Order	Number of Group			
1	1		11	1
2	1		12	5
3	1		13	1
4	2		14	2
5	1		15	1
6	2		16	14
7	1		17	1
8	5		18	5
9	2		19	1
10	2		20	5

THE ENGLISH HANDBOOK OF GRAMMAR, STYLE, AND COMPOSITION

- This book illustrates the rules and numerous exceptions that are characteristic of the English language, in great depth, detail, and clarity.

- Over 2,000 examples comparing correct and wrong usage in all areas of grammar and writing.

- Solves the usual confusion about punctuation.

- Illustrates spelling "tricks" and how to remember correct spelling.

- Teaches how to acquire good writing skills.

- Provides special learning exercises at the end of each chapter to prepare for homework and exams.

- Fully indexed for locating specific topics rapidly.

Available at your local bookstore or order directly from us by sending in coupon below

---

RESEARCH and EDUCATION ASSOCIATION
61 Ethel Road W., Piscataway, New Jersey 08854
Phone: (201) 819-8880

Charge Card Number

☐ Payment enclosed

Please check one box:
☐ Visa
☐ Master Card

VISA   MasterCard

Expiration Date ____ / ____
                 Mo   Yr

Please ship the "English Handbook" @ $12.95 plus $4.00 for shipping.

Name _____

Address _____

City _____ State _____ Zip _____

THE BEST AND MOST COMPREHENSIVE IN TEST PREPARATION

GRE
ECONOMICS TEST

GRADUATE RECORD EXAMINATION

- Based on the most recent exams.

- *SIX* full length exams. Over 650 pages.

- Each exam is 2 hours and 50 minutes.

- Explanations to questions are extensively illustrated in detail.

- Almost every type of question that can be expected on the Economics Test.

- Complete Answer Key follows every exam.

- Enables students to discover their strengths and weaknesses and thereby become better prepared.

Available at your local bookstore or order directly from us by sending in coupon below.

RESEARCH and EDUCATION ASSOCIATION
61 Ethel Road W., Piscataway, New Jersey 08854
Phone: (201) 819-8880

Charge Card Number

Please check one box:
- ☐ Payment enclosed
- ☐ Visa
- ☐ Master Card

Expiration Date ____ / ____
              Mo   Yr

Please ship the "GRE Economics Test" @ $15.95 plus $4.00 for shipping.

Name _____

Address _____

City _____ State _____ Zip _____

HANDBOOK AND GUIDE FOR SELECTING A CAREER

AND PREPARING FOR THE FUTURE

For:

- Young Job-Seekers
- Persons Seeking a Career Change
- Persons Entering the Labor Force Later In Life

NEW 1989-90 EDITION

Over 250 careers are covered. Each career is described in detail including:

- Training and Education
- Character of the Work Performed
- Working Conditions
- Amount of Earnings
- Advancement Opportunities

Available at your local bookstore or order directly from us by sending in coupon below.

RESEARCH and EDUCATION ASSOCIATION
61 Ethel Road W., Piscataway, New Jersey 08854
Phone: (201) 819-8880

VISA   **MasterCard**

Charge Card Number

Please check one box:
- ☐ Payment enclosed
- ☐ Visa
- ☐ Master Card

Expiration Date _____ / _____
              Mo   Yr

Please ship the "Career Handbook" @ $15.95 plus $4.00 for shipping.

Name _____

Address _____

City _____ State _____ Zip _____

HANDBOOK AND GUIDE FOR
COMPARING and SELECTING
COMPUTER LANGUAGES

BASIC	PL/1
FORTRAN	APL
PASCAL	ALGOL-60
COBOL	C

- This book is the first of its kind ever produced in computer science.

- It examines and highlights the differences and similarities among the eight most widely used computer languages.

- A practical guide for selecting the most appropriate programming language for any given task.

- Sample programs in all eight languages are written and compared side-by-side. Their merits are analyzed and evaluated.

- Comprehensive glossary of computer terms.

Available at your local bookstore or order directly from us by sending in coupon below.

RESEARCH and EDUCATION ASSOCIATION
61 Ethel Road W., Piscataway, New Jersey 08854
Phone: (201) 819-8880

VISA   MasterCard

Charge Card Number

Please check one box:
- ☐ Payment enclosed
- ☐ Visa
- ☐ Master Card

Expiration Date ____ / ____
          Mo   Yr

Please ship the "Computer Languages Handbook" @ $8.95 plus $2.00 for shipping.

Name _____

Address _____

City _____ State _____ Zip _____

HANDBOOK OF MATHEMATICAL, SCIENTIFIC, and ENGINEERING FORMULAS, TABLES, FUNCTIONS, GRAPHS, TRANSFORMS

A particularly useful reference for those in math, science, engineering and other technical fields. Includes the most-often used formulas, tables, transforms, functions, and graphs which are needed as tools in solving problems. The entire field of special functions is also covered. A large amount of scientific data which is often of interest to scientists and engineers has been included.

Available at your local bookstore or order directly from us by sending in coupon below.

RESEARCH and EDUCATION ASSOCIATION
61 Ethel Road W., Piscataway, New Jersey 08854
Phone: (201) 819-8880

Charge Card Number

Please check one box:
☐ Payment enclosed
☐ Visa
☐ Master Card

Expiration Date ____ / ____
                Mo   Yr

Please ship the "Math Handbook" @ $24.85 plus $4.00 for shipping.

Name _____

Address _____

City _____ State _____ Zip _____

THE BEST AND MOST COMPREHENSIVE IN TEST PREPARATION

GRE
MATHEMATICS TEST
GRADUATE RECORD EXAMINATION

- Based on the most recent exams.
- *SIX* full length exams.
- Each exam is 2 hours and 50 minutes.
- Explanations to questions are extensively illustrated in detail.
- Almost every type of question that can be expected on the Mathematics Test.
- Complete Answer Key follows every exam.
- Enables students to discover their strengths and weaknesses and thereby become better prepared.

Available at your local bookstore or order directly from us by sending in coupon below.

RESEARCH and EDUCATION ASSOCIATION
61 Ethel Road W., Piscataway, New Jersey 08854
Phone: (201) 819-8880

VISA   MasterCard

Charge Card Number

Please check one box:
- ☐ Payment enclosed
- ☐ Visa
- ☐ Master Card

Expiration Date ____ / ____
                 Mo   Yr

Please ship the "GRE Mathematics Test" @ $15.95 plus $4.00 for shipping.

Name _____

Address _____

City _____ State _____ Zip _____

THE PROBLEM SOLVERS

The "PROBLEM SOLVERS" are comprehensive supplemental textbooks designed to save time in finding solutions to problems. Each "PROBLEM SOLVER" is the first of its kind ever produced in its field. It is the product of a massive effort to illustrate almost any imaginable problem in exceptional depth, detail, and clarity. Each problem is worked out in detail with step-by-step solution, and the problems are arranged in order of complexity from elementary to advanced. Each book is fully indexed for locating problems rapidly.

ADVANCED CALCULUS
ALGEBRA & TRIGONOMETRY
AUTOMATIC CONTROL
   SYSTEMS/ROBOTICS
BIOLOGY
BUSINESS, MANAGEMENT,
  && FINANCE
CALCULUS
CHEMISTRY
COMPLEX VARIABLES
COMPUTER SCIENCE
DIFFERENTIAL EQUATIONS
ECONOMICS
ELECTRICAL MACHINES
ELECTRIC CIRCUITS
ELECTROMAGNETICS
ELECTRONIC COMMUNICATIONS
ELECTRONICS
FINITE & DISCRETE MATH
FLUID MECHANICS/DYNAMICS
GENETICS

GEOMETRY:
PLANE • SOLID • ANALYTIC
HEAT TRANSFER
LINEAR ALGEBRA
MACHINE DESIGN
MECHANICS: STATICS • DYNAMICS
NUMERICAL ANALYSIS
OPERATIONS RESEARCH
OPTICS
ORGANIC CHEMISTRY
PHYSICAL CHEMISTRY
PHYSICS
PRE-CALCULUS
PSYCHOLOGY
STATISTICS
STRENGTH OF MATERIALS &
   MECHANICS OF SOLIDS
TECHNICAL DESIGN GRAPHICS
THERMODYNAMICS
TRANSPORT PHENOMENA:
MOMENTUM • ENERGY • MASS
VECTOR ANALYSIS

If you would like more information about any of these books, complete the coupon below and return it to us or go to your local bookstore.

RESEARCH and EDUCATION ASSOCIATION
61 Ethel Road W. • Piscataway • New Jersey 08854
Phone: (201) 819-8880

Please send me more information about your Problem Solver Books

Name _____

Address _____

City _____ State _____ Zip _____